An Introduction to Planetary Defense

A Study of Modern Warfare Applied to Extra-Terrestrial Invasion

Other Books by Travis S. Taylor

Warp Speed

The Quantum Connection

Von Neumann's War (and John Ringo)

The Appendix to Deep Space Probes 2nd Edition (by Greg Matloff)

Neighborhood Watch Final Report (ebook at www.baen.com)

Coming in 2007

The Tau Ceti Agenda

Vorpal Blade (and John Ringo)

An Introduction to Planetary Defense

A Study of Modern Warfare Applied to Extra-Terrestrial Invasion

Travis S. Taylor
Bob Boan

with contributing authors

R. Charles Anding
T. Conley Powell

BrownWalker Press
Boca Raton • 2006

An Introduction to Planetary Defense:
A Study of Modern Warfare Applied to Extra-Terrestrial Invasion

The cover art includes an image of the Earth and Moon compiled from NASA/NOAA/USGS data. The image of the Earth/Moon was by Stockli, Nelson, Hasler; Laboratory for Atmospheres, Goddard Space Flight Center, http://rsd.gsfc.nasa.gov/rsd.

BrownWalker Press
Boca Raton , Florida
USA • 2006

ISBN: 1-58112-447-3 (paper)
ISBN: 1-58112-448-1 (ebook)

BrownWalker.com

Dedication

We dedicate this text to those unfortunate citizens and heroes who lost their lives in the September 11, 2001 terrorist attacks on America. America was unprepared for those attacks. It is our hope and desire that with this book we will stimulate enough impetus to take advantage of lessons learned from our lack of planning and preparation. Being similarly unprepared for an invasion from extra-terrestrials would prove far more catastrophic and would likely end humanity. A response to the call to plan and prepare will add great purpose to the September 11, 2001 victims' unfortunate fates.

Contents

About the Authors

Principle Authors:

Dr. Travis S. Taylor

Travis Shane Taylor is a born and bred southerner and resides just outside Huntsville, Alabama. He has a Doctorate in Optical Science and Engineering, a Master's degree in Physics, a Master's degree in Aerospace Engineering, all from the University of Alabama in Huntsville; a Master's degree in Astronomy from the Univ. of Western Sydney, and a Bachelor's degree in Electrical Engineering from Auburn University. He is a licensed Professional Engineer in the state of Alabama.

Dr. Taylor has worked on various programs for the Department of Defense and NASA for the past sixteen years. He is currently working on several advanced propulsion concepts, very large space telescopes, space based beamed energy systems, future combat technologies and systems, and next generation space launch concepts. He is also involved with multiple MASINT, SIGINT, IMINT, and HUMINT concept studies.

He has published over 25 papers and the appendix on solar sailing in the 2nd edition of *Deep Space Probes* by Greg Matloff.

His first science fiction novel is, *Warp Speed*, and his second is *The Quantum Connection* published by Baen Publishing. He is also working on two different series with best-selling author John Ringo also by Baen Publishing. He has several other works of both fiction and nonfiction ongoing.

Travis is also a Black Belt martial artist, a private pilot, a SCUBA diver, races mountain and road bikes, competed in triathlons, and has been the lead singer and rhythm guitarist of several hard rock bands. He currently lives with his wife Karen, his daughter Kalista Jade, two dogs Stevie and Wesker, and his cat Kuro.

Dr. Bob Boan

Dr. Boan has been an active member of the space community for over a quarter of a century. He has worked on a variety of manned and unmanned space programs at different levels of responsibility over that time. Prior to his space experience he was a member of academia. He taught primarily chemistry. He taught courses from high school through graduate school in several states.

Dr. Boan is also recognized as a community expert on SIGINT, IMINT, and Communication systems and concepts. He also has significant MASINT experience. He has multiple relevant patents and technical publications.

Dr. Boan has attended a variety of colleges and universities. He received his BS from Campbell University, then Campbell College. His Master's was awarded by the University of Mississippi. He earned his doctorate at the Florida Institute of Technology.

Dr. Boan has significant experience as an author. He has authored a number of technical publications. He has compiled a book of witticisms and a fiction novel and is currently working on two new novels; one of which is science fiction he is coauthoring with "Doc" Travis Taylor.

Dr. Boan has been featured in Marquis' *Who's Who in the South and Southwest* and in Marquis' *Who's Who in America.*

Contributing Authors:

Mr. R. Charles Anding

Mr. Anding received his bachelor's degree in Electrical Engineering from Mississippi State University. He has additional studies in systems engineering, digital signal processing and electromagnetic environmental effects. Mr. Anding has applied his creativity and expertise to solve a diversity of engineering problems for over 25 years. He has designed electronics and systems for space, military, industrial and medical products. He was the prime contractor's chief engineer for the design and development of a furnace system to grow semiconductor crystals in microgravity on both the Space Shuttle and the International Space Station. He supported its use on multiple Spacelab missions, including training of the astronauts and sitting console for payload operations during missions.

Mr. Anding was the chief engineer, along with Dr. Taylor as chief scientist, for the development of a novel new mission and spacecraft for exploring Pluto. He has designed and supported equipment on Navy fighter aircraft, Army main battle tanks, and attack helicopters. Non-invasive cardiac monitors for medical market, industrial robotics for the nuclear segment and user authorized handguns are just a few more examples of his broad experience base. He is currently designing controls for demilitarization of binary chemical weapons and beginning research and development for future fuel cell based power systems for rugged environments as well as applying unmanned aerial vehicles for defense purposes.

Dr. Thomas Conley Powell

CONLEY POWELL holds a B.A. in physics from Berea College, an M.S. in engineering science from the University of Tennessee Space Institute, and a Ph.D. in mechanical engineering from the University of Kentucky. He is a senior scientist with BAE Systems in Huntsville, Alabama. Before joining BAE, he was a faculty member at the Space Institute; a member of the technical staff at Arnold Engineering Development Center, near Tullahoma, Tennessee; and a member of the technical staff of Teledyne Brown Engineering, in Huntsville. He has taught graduate courses in subjects ranging from astrophysics to nuclear engineering, and has worked in areas as diverse as aircraft control and nuclear fusion. However, his specialties are space trajectories, attitude dynamics, and numerical analysis. Recently he has developed an innovative fire-control system for artillery and surface-to-surface and surface-to-space rockets. He is writing a textbook on orbital mechanics.

He lives in Athens, Alabama with his wife Judy, who is an interior designer, and their two cat, Mr. P. and two dogs, Bonnie and Scruffy MacGruff. His hobbies (in no particular order) are science fiction, crackpot literature, and weight lifting.

Introduction

This book was written to be considered seriously as a starting point for developing defensive and offensive concepts in the event that an attack from advanced extra-terrestrials occurs. The goal of this textbook is not to refute, discuss, defend, or even enter into an argument about government conspiracies, UFO cover-ups, alien autopsies, or any other examples of the "UFOlogy" genre. While we call upon science fiction for terms and examples, this text is actual science and experience. We base all our analyses on the facts to the extent possible. We supplement these facts with logical extension where possible.

It is our opinion that the possibility, however small or large, of such an attack exists from the simple fact that there are numerous other star systems in addition to ours. The possibility of extra-terrestrial life is, without a doubt, finite. In fact, the probability is overwhelming. We would be extremely arrogant, and perhaps ignorant, to believe Earth to be the only celestial body with a lifeform somewhat similar to our own. Whether or not the nonzero probability is large or small is of course unknown through scientifically verifiable and widely accepted evidence. However, the probability is finite! This being the case, it is also likely that if these ETs exist, some of them might wish us harm in some fashion or other for reasons known only to them.

We have several goals for this book. Those goals include:

- to discuss briefly the probability of a visit by ET
- to discuss the possible types of these ETs
- to categorize ETs by their ranges of power
- to take a first look at weapons for offense and defense
- to take a first look at the possible intentions for visits by the different categories of ETs
- to issue a call for preparation for a visit from ET
- and depending on their respective degree of power, could we defend ourselves against them or not? This is a question we hope to at least begin to answer - if not answer, begin thinking about how to answer.

A text of this type might have been very useful to the Native Americans when the Europeans – though not from a distant celestial body, ETs of a sort themselves - entered the Americas. This or any other similar text would have been of value only if the Native Americans had heeded the call for preparation as we ask our civilization to do. The analogy holds well in that the Europeans were from a much more technologically advanced civilization than the Native Americans – just as a visiting ET would be more advanced than we are. The Europeans entered the Native society and began to conquer and assimilate it. In some cases this occurred peacefully and in others, not so peacefully. There was little that the natives could do to stop the advance of the White Man. The best they could do was delay the inevitable even though they slowly adopted the military technology of the Europeans – it was too little, too late. However, as we will discuss later in the book, the Native Americans employed many tactics that are today known as "guerilla" tactics or more properly asymmetric warfare.

Asymmetric warfare is an overpowering in numbers and/or technologically advanced force attacking a smaller – much smaller in most cases – force. An example of this type of warfare is the second Gulf War or Operation Iraqi Freedom whereas the superior forces of the United States Coalition Force were plagued by the Iraqi insurgent forces. A small number of insurgents employed improvised explosive devices and other asymmetric warfare tactics to prolong their presence on that culture. In an alien invasion humanity might just find itself in the situation of the smaller "insurgent" force.

We would be the Natives if attacked by aliens. The human race is to be considered as one civilization with multiple factions within it. Again the Native American analogy holds true since there were also many various Indian Nations. The Europeans, ETs this time, are coming we must assume. Will they be nice or nasty? Do we wish to become extinct as some Natives did? Do we wish to travel our own "Trail of Tears?" Do we wish to be assimilated into a European society as happened to some of the Native Americans? Or will we fight back? Perhaps we should prepare for the worst and hope for the best. Now is the time for preparation before it is too late!

With the data and discussion gathered within this book, we hope to spark some interest in developing a defense strategy for the human civilization. In the event that one day some smooth talking aliens show up with some shiny trinkets and strands of beads, hopefully, we will be more prepared than the Native Americans were. If ET comes will we trade Manhattan, the US or the Globe? If the shiny beads are a cure for cancer or world hunger or perpetual energy, what would we trade? It would be too late at that point to have much of a bartering position if the aliens could simply take what they wanted.

It is said that those who do not learn from history are doomed to repeat it. We should learn from the fate of the American Indian. We should develop a strategy for dealing with the arrival of ET. It is likely that this strategy is best developed as a people of the planet as opposed to that of a country. Had the American Indian had the communications and transportation capabilities that we enjoy today, they could have joined all of the Indian nation forces to produce a force to produce an entity that may have immediately defeated the European ETs despite having inferior military technology. Sheer numbers and knowledge of the battle fields would have overwhelmed the Europeans and sent them home unlikely to return, at least without significant preparation.

Had the Indians themselves been making preparations for their return and had they joined forces, history would have a very different story to tell. They probably would have defeated the intruders even with their inferior weapons. On the other hand, as we will discuss in the various sections of this book, cooperative preparation may not be very easy to achieve. Therefore, the USA should at least do its part whether the remainder of the countries of the Globe participate or not.

If a hostile ET showed up and left, we would prepare for his return. He, too, would prepare for his return with more forces and/or technologies. We should prepare now and improve our odds that he would not return. Let's PREPARE NOW AND SURVIVE LATER. This is our thrust in this book. Being prepared will continue to percolate as the best strategy throughout the discussion in this study.

We are merely suggesting the development of a strategy. A strategy is needed even if it is never put into action. We have a responsibility to ourselves and our progeny to prepare for the possible. We leave the implementation of the strategy to our elected officials. We will however attempt to define levels of power at which selected aspects of a strategy are rendered futile.

A major issue in convincing governments and the public that we need to prepare is that we do not know whether or not we are alone in the universe. If we are the only intelligent lifeform, an interesting question relative to our existence arises. Are we an accident? Are we the result of an experiment? If we were an accident, then we weren't intended to exist at all. Certainly if we were not intended to exist it is logical to assume that no intelligent lifeform was intended to exist. Were we an unintended and unexpected byproduct of evolution? If we are the result of experiment, then the experiment must have had a bad result. If the result were not bad, why would there not be other intelligent lifeforms in the universe? If either of the scenarios is true, then it is reasonable to assume that we are the only intelligent lifeform in the universe. There are of course religious arguments for a single species universe, but the science is weak and will not be discussed here.

If we are the only intelligent lifeform in the universe, why is there such a vast universe? It is certainly not obvious what benefits we get from the rest of the universe if any. It is also not obvious that the rest of the universe takes anything beneficial from us. Could so many stars and celestial bodies be required to stabilize our solar system so that we select lifeforms are allowed to exist? One can only reach the conclusion that if we are the only intelligent lifeform, we are incredibly special. But, why us? If one scrutinizes the history of mankind, one would be hard pressed to find the type of universal positive behavior that one would anticipate and expect from a very special, favored species. How would we explain Hitler, Genghis Kahn, Idi Amin, Saddam Hussein, the Khmer Rouge and other sadistic rulers in this light? Each of these people probably meant well by their standards but not by the standards of the mainstream populace.

Both accident and bad experiment as explanations for our being alone in the universe are in contradiction with popular beliefs. A fundamental belief in many religions is that there is an omnipotent, benevolent Supreme Being who is the creator. If the Supreme Being is truly omnipotent, there could not be an accident. There could not be a result that is unexpected, only ones that are predetermined. An experiment could not go awry. An experiment could not have a bad ending except by design. A bad ending would be inconsistent with the benevolence of this so-called creator. The outcome would be well known at the beginning.

So, we suggest that we continue along the lines of reasoning and the more scientifically supported concept that we are not alone in the universe. How much company we have is unknown, although we will discuss this later. The point is that we most likely are not alone. At least logic would suggest such a conclusion. Therefore we feel that we should be prepared for future interactions with other civilizations. If we do not prepare now, it is likely that we will pay for our present inactivity in the future with our extinction.

Finally, the question arises: "How do we prepare?" We will discuss in this book various concepts of warfare, force-on-force theory, and possible future technological and military capabilities. Also, in the light of many of the modern natural disasters, it is quite clear that the civilian aspect of humanity has little, if any, emergency response capabilities that could be implemented effectively in a planetary wide disaster such as a planetary wide invasion (or even large scale meteor impacts). We attempt to address the initial aspects of modern military and civil defense requirements for surviving an invasion from advanced alien civilizations. We are making no claim that this is an all encompassing text on the subject. In fact, this book is merely scratching the surface of the topic and we hope that we can continue to dig deeper into the new aspects and paradigms necessary for planetary defense. Hopefully future editions and even *Advanced Planetary Defense Concepts* with more detailed applications and scenario simulations will be developed in the future. In the meantime, we will begin with ***An Introduction to Planetary Defense***.

1

The Statistics – Probability of an Alien Invasion

Is an alien attack possible? Of course it is. Statistically speaking, almost anything is possible. There is a better question to ask, which is, "What is the probability of an alien attack"? The likelihood of an alien attack can be discussed statistically but only after making some logical assumptions. For now, we will take an elementary mathematics approach to calculating this probability. This is an incomplete approach but is a good place to start for demonstration purposes.

There are about four hundred billion stars in the Milky Way galaxy alone. We know that at least one star system (our own) within the Milky Way Galaxy has developed intelligent life, at least as defined by our scale. So that suggests statistics of at least one civilization per galaxy. How many galaxies are there? There are billions, maybe more. So, there should be billions of star systems with intelligent civilizations even if there is only one intelligent civilization per galaxy!

Just one inhabited star system per galaxy seems preposterously low though. Again, there are four hundred billion stars in the Milky Way. That leads us to a probability of 0.25×10^{-11} of intelligent life existing. Such a probability would lead us to the conclusion that we are an accident or a special, unique freak of nature. Perhaps a more likely approximation can be gathered by making some further assumptions.

1.1 The Drake Equation

How many planets are there in the galaxy which host technical civilizations? This question has been pondered by mankind since before the beginning of our own technical civilization, which is only a few hundred to a few thousand years old depending upon the definition used. There is no doubt that primitive man must have wondered what amazing or horrible things could be lurking out in the stars. In 1961 the Cornell astronomer Frank Drake developed an approach for estimating the number of civilizations in the Milky Way. Drake proposed what has since become known as the "Drake Equation". The approach has become generally accepted in the Search for Extra-Terrestrial Intelligence

(SETI) community as the tool for analyzing factors that are crucial to civilization development. The equation is

$$N = R_* f_p n_e f_l f_i f_c L \tag{1.1}$$

where N is the number of civilizations in the Milky Way that developed systems which produce created electromagnetic emissions detectable from Earth, R_* is the rate of formation of stars suitable for the development of intelligent life, f_p is the probability of those stars having planetary systems, n_e is the number of planets per solar system that can support life, f_l is the probability that life actually occurs on those planets, f_i is the probability that intelligent life occurs, f_c is the probability that the intelligent life develops technology that produces electromagnetic signals detectable from Earth, and L is the length of time that the civilization is sending out these detectable signals. A discussion of each of these factors is in order.

R_* is fairly well understood. Observations from ground and space based telescopes over the last few decades have enabled a good estimate of this value to be about 1.5; that is on average 1.5 new stars per year. As we know, half stars are not created. We would think that about 3 new stars that are suitable for life to develop around are being formed every two years. So,

$$R_* = 1.5 \tag{1.2}$$

Exciting new observations of many planets around many stars leads us to understand that planets are a byproduct of star formation. So, as a star forms, planets are frequently born as a byproduct. There are some situations where this appears not to be the case. But these instances are exceptions, not the rule. Modern astronomical observations suggest that about ninety percent of stars formed produce planets. This leads to a value of 0.9 for f_p.

$$f_p = 0.9 \tag{1.3}$$

The number of planets per solar system that are suitable for life (i.e., number of "Earthlike" planets) is more considerably difficult to estimate. Some astronomers believe that these types of planets must exist within the zone around their star where liquid water can exist. We refer to water as the liquid state of the compound H_2O as opposed to steam or ice as the gaseous and solid states, respectively. Until recently, it was not realized that the Jovian moon Europa has a very deep liquid water ocean beneath its icy surface. Also, at one time Mars must have had liquid water on its surface. This leads to an interesting discussion about the potential for life on Europa and Mars. We should be prepared for the fact that if there is life on Europa, it might be anatomically dissimilar to lifeforms with which we are familiar. The lifeforms, including intelligent ones, might be so dissimilar that we might not immediately identify them as intelligent. An important question is "Why must the life be water based?" Titan, for example, might very well have liquid methane oceans, lakes, or rivers. It is possible that methane based lifeforms could and do exit. However, to keep the discussion simpler, we will only consider water-based life at this point as we have a better chance of understanding such lifeforms.

Using the elementary approach, we could say that at least three out of the 15 or more planet sized objects in our solar system have or have had liquid water on or in them at some point in their lifespans. Therefore, the probability (again from the freshman approach) that a planet in our solar system has water is three divided by 15 or one fifth (0.2). And the number of planets is three. If other than water based life is considered we would anticipate that the number becomes larger; the probability will vary with the other mediums assumed in addition to water. Again, for this argument we will only consider water-based lifeforms. Also, it is unknown if our solar system is typical although we are beginning to observe many star systems with multiple planets. Since Mars and Europa are not completely Earthlike we will be conservative in our estimates and will divide the value by two. So the number of Earthlike planets or planetoids per solar system is

$$n_e = 1.5 . \tag{1.4}$$

For convenience we will use "planet" to refer to planets, planetoids, moons or other bodies which could be a home to ET. Just because a given planet is suitable for life does not necessarily mean that it will develop life. Or does it? Scientists have found strange forms of life in the most hostile environments on Earth. Chemosynthesis based creatures have been found in underwater caves and at the bottom of the sea. Laboratory experiments show that amino acids can form out of a chemical soup as a result of the introduction of electric arcs. These amino acids are the building blocks of life as we know it and much more highly energetic conditions can occur from planetary electric storms than those experienced in chemical soup tests. These more energetic conditions increase the probability of biogenesis. It has become widely accepted that if a planet can support life in any weird and imaginable way, then it will, eventually, almost certainly, form life. The SETI community typically uses a value of one for the fraction of planets that actually develop life. Therefore

$$f_l = 1 . \tag{1.5}$$

Of the number of planets that develop life, how many will form intelligent life? Originally, Drake and his colleagues decided that all planets that form life would eventually develop intelligent life. Mars may, or may not, very well be in contradiction of this conclusion. Planetary scientists using probes have recently seen evidence of freely flowing water on the Martian surface having existed sometime in the past million years or so. It is very likely that eventually we will find microbial and other simple life on Mars. There is also the theory introduced by NASA that microbial life did exist on Mars based on evidence from meteor rocks found on Earth. If this is the case and Mars did form life, where is the intelligent life? Perhaps it is extinct? Perhaps it is there and we fail to recognize it? Perhaps it has yet to develop? Perhaps its evolution was ended by some cataclysmic phenomena.

At this point in the Martian evolutionary cycle, all evidence points to Mars maturing and aging. The Martian environment is very harsh and it would appear that Mars is not evolving life as we recognize it any longer. Therefore, it is possible that the original conclusion by Drake and his colleagues was a bit optimistic. Of course, there are those who might claim that a civilization could have formed on Mars and we just never knew it.

That is possible. However, it would be difficult for such a civilization to leave without a trace unless it was very, very long ago and has been decayed and engulfed by the surface dust over time. Thus far we have found no evidence of such. For now we will conclude that Mars has never evolved intelligent life.

What about Europa? With its deep ocean and most likely very volcanic sea bottom due to the huge gravitational tides from Jupiter and its other moons, it is highly probable that there are strange little fish swimming around in its oceans right now. We can make no further conclusions about Europa simply because we have a severe lack of hard data. What we can assume is that Europa is at the beginning of its evolution and that intelligent sea-based creatures may still develop there eons from now.

So, from the three planets in our solar system that have most likely created life, presently only one of them has appeared to have developed intelligent life. It is still possible that Europa will. It is highly unlikely that Mars will. Of course, it is possible that a string of comets could strike Mars, heat it up, and give it a thicker atmosphere and an environment more suitable for evolution of life. We will not consider these possibilities whose probabilities are too difficult to calculate accurately at this time due to our lack of a priori knowledge of each event should they happen at all.

It is possible that as the Solar System matures, Venus and/or later Mercury may develop an aqueous environment. If that were to happen, they too might be capable of supporting life – even intelligent life as we recognize it. This possibility is also left out of our statistical treatment of the possibility of intelligent life other than that which we recognize on Earth.

Therefore, the elementary approach suggests that two thirds of the planets in our solar system that can develop life will develop intelligent life. This means that

$$f_i = \frac{2}{3} \approx 0.67 \ . \qquad (1.6)$$

Is it possible that any of these planets, which develop intelligent life, will develop civilizations with the capability to communicate with other planets? Assume now that communication merely means that we can detect some form of electromagnetic signal from their civilization. Drake's group estimated this number to be between ten and twenty percent of the intelligent civilizations. This range of percentages seems far too low and far too conservative. If a civilization develops, eventually they can be detected as our communications technology catches up to, or surpasses, theirs. It is reasonable to assume that there are lifeforms that communicate in spectral bands that we have yet to discover or know how to use. It is also reasonable to assume that there are lifeforms that use communication technologies beyond our understanding of physics and therefore undetectable by our technologies. The problem here with these conservative estimates is that Drake used a far too loose a definition for "civilization" or for "intelligent life". One of these "loose" assumptions was that dolphins are intelligent but never decided to build communication equipment. Nonsense! Dolphins are not intelligent life. Agreed they are smarter than the average bear, but they do not build tools and have not, to our knowledge, made attempts to communicate with the human race on a mass or any other basis. Using this assumption, dogs, chimps, parrots, cats, and a lot of other creatures would be considered intelligent. Are we just too dumb or just not advanced enough to understand these species? Most of us would argue definitely not, especially when referring to dogs and cats!

We will define here that an intelligent civilization is a civilization that develops a society, tools, and the desire to improve conditions through use of the race's abilities. Perhaps many of the creatures on Earth will eventually evolve to become "intelligent" but for now only humans have exhibited the traits as per our definition.

It is possible however that some of these civilizations are smarter and more xenophobic than we are. The more xenophobic-minded civilizations might be more protective of their society and will hide their communications signals with hopes of not being detected by hungry aliens. This could be an excellent defensive strategy. Drake did suggest there would be such civilizations.

As the world witnessed during the Cold War, secrets are hard to keep. If some industrious group really desires to learn a secret, they will eventually figure out a method to do so. A really good example of why a civilization cannot hide easily from advanced civilizations is the development of extremely large aperture telescopes. We are not talking about slightly larger than Hubble Space Telescopes here. We are talking about telescopes with primary apertures hundreds of kilometers in diameter. NASA is already conducting studies and experiments to develop such telescopes that could even image features on the extra-solar planet's surfaces, not just images of distant planets themselves. Recently in his book *The Sun as a Gravitational Lens: Proposed Space Missions*, Claudio Maccone, a space scientist at Alenia Aerospazio in Turin, Italy, suggests implementing the sun's gravitational lensing effect (as predicted by Einstein's General Theory of Relativity) in a telescope. In other words, the sun could be used as a lens producing an aperture the diameter of the sun. Its focus is at about six hundred astronomical units away. If the proper equipment were placed at the Solar focus, extremely detailed images of extra-solar planetary surfaces could be captured. Civilizations only moderately more advanced than we are could quite possibly implement such a telescope. Therefore, even quiet civilizations could be seen through such a telescope. Eventually, detection of such civilizations may not depend on their directly communicating with us by sending signals. The light from their star bouncing off of their alien rooftops and then onward to our giant telescope may be enough.

It is most likely that any advanced civilization could eventually detect any lesser-advanced civilization. Therefore, Drake's estimates of ten to twenty percent seem quite low indeed! Barring a major planet-killing event all "intelligent civilizations" per definition will eventually be detectable by a more advanced civilization. Hence, the value of the communication (which should be changed to detection) fraction is one hundred percent of all lesser-advanced civilizations. Making the arrogant assumption that half of all civilizations are less advanced than we are suggests that fifty percent of all civilization are detectable. For conservatism we will assume that only a quarter of those civilizations are less advanced than us, resulting in

$$f_c = 0.25 \tag{1.7}$$

This ironically is not far from Drake's original conclusion but for different reasons.

The last factor pertaining to how long will a communicating or detectable civilization remain detectable is really a guess at best. Drake suggested between one thousand and a hundred million years. Again, he used different criteria for communication. If we assume detection as part of communication, then any lesser-advanced civilization would remain detectable as long as they existed or until they leapfrogged us in technological advancement or lost the resource base to sustain their communication technology. For example, if a civilization had ever developed on Mars, we would see its ruins with our telescopes even if the civilization no longer existed for some reason. This is true unless it were a totally subterranean culture. The length of time that the Martian civilization would be detectable would be the length of time that the Martian civilization overlaps the length our civilization lifetime. This could only be a few million years! OUR SPECIES ANCESTORS ARE ONLY ABOUT 3 TO 5 MILLION YEARS OLD – the telegraph is less than 175 years old.

We have produced electromagnetic spectra potentially detectable by ET for less than 100 years by technical means such as those we know today. Our earliest signals have yet to reach many planets with the potential to host intelligent life. We may not have been detected by ET yet because we are too immature. However, if we consider the possibility of our speech or consciousness being detectable by more advanced technology than we possess, we could have been detectable for 3-5 million years (although the human species is about 200,000 years old its ancestral lineage is much older).

Although the likelihood that the detectable time of alien civilization might be very large, some conservatism should be used in order to have a plausible and realistic answer. Assume that this factor relates to the detection of live civilizations and to the overlap of that civilization being alive now while we are looking. This assumption reins in the range of values. Since we are missing a validated method to determine this value, we will use Drake's most conservative predicted value, where

$$L = 1000\,. \qquad (1.8)$$

Inserting the values for the above factors into equation 1.1 yields a value of about

$$N \approx 340 \qquad (1.9)$$

detectable civilizations during the life of our civilization. The application of more conservative or liberal values for any or all of the factors changes the results significantly. With unknown factors chosen mostly by assumption, it is probably better to discuss a range of values of N in orders of magnitude. Assuming the lower end of the range to be an order of magnitude less and the upper range to be an order of magnitude higher then N is somewhere between thirty-four and three thousand four hundred detectable civilizations!

Now consider equation 1.1 without the factor L. This gives a rate of detectable civilizations developed per year as

$$dN = R_* f_p n_e f_l f_i f_c = \frac{340}{1000\text{years}} = 0.340\,. \qquad (1.10)$$

This set of mathematics leads us to the conclusion that a new detectable civilization develops roughly every three years. Remembering the order of magnitude error bars suggests a range on the rate at which detectable civilizations develop from once every four months to once every 30 years. Therefore, there is a reasonable probability of at least two new intelligent civilizations developing during the average life time of a human being. If we assume the development of new intelligent civilizations to have been constant at one every thirty years, at least 100,000 detectable civilizations would have developed during man's three to five million years on Earth. For further discussion, we will consider only the median value of one civilization every three years. However, be sure to remember the impact of the possible range of values.

During the time taken for man to mature from the point of standing upright to building the first telescope (roughly 3-5 million years), more than one million detectable civilizations could have possibly developed. Even if we use the lower end of estimates of the age of the Earth, three billion years, more than one billion detectable civilizations have possibly developed since Earth was formed! Figure 1.1 shows a log-log scale graph of the number of civilization that develops as a function of time in years. The range is from one year to five billion years. Note that in the 10 thousand years of so-called human civilization more than fifty thousand detectable civilizations may have formed. In the hundred or so years that humans have mastered flight more than 50 alien civilizations may have developed.

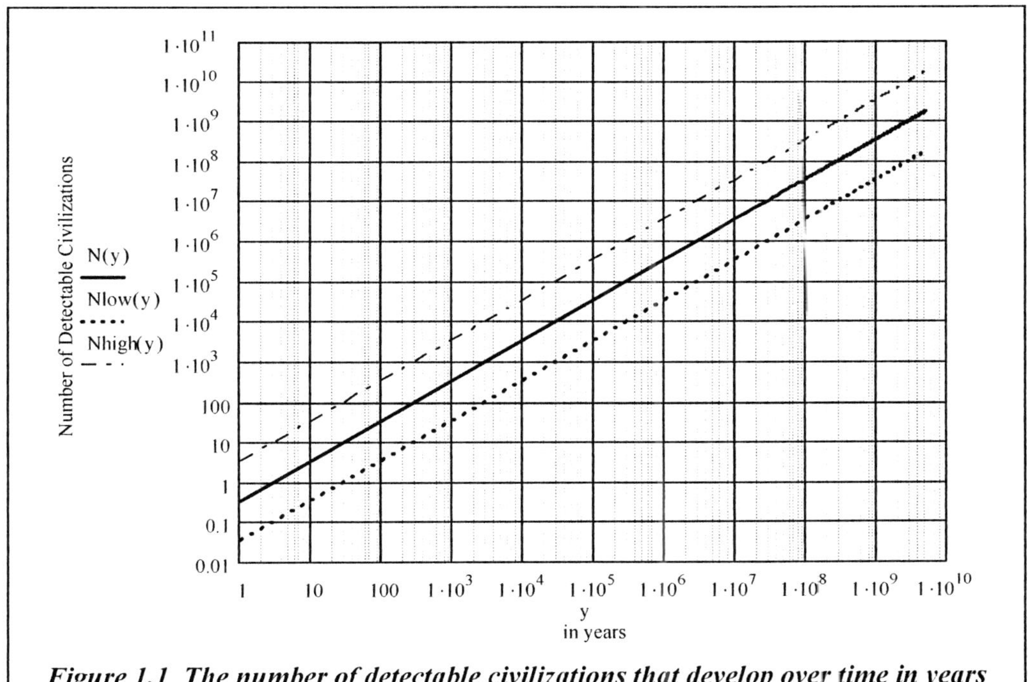

Figure 1.1 The number of detectable civilizations that develop over time in years

Now consider equation 1.10 as the rate of development of new detectable civilizations. The argument made previously for the value of f_c was that 25 – 50% of the civilizations were lesser-advanced. In other words, 50 -75% of the intelligent civilizations were more advanced than ours. This also suggests that equation 1.10 is also the rate of development of more-advanced civilizations that are undetectable to us for whatever reason.

Assuming that some fraction of these advanced cultures decide to travel to other systems then the number of "travelers" can be expressed as

$$dN = R_* f_p n_e f_l f_i f_c f_t \tag{1.11}$$

where f_t is the percentage of traveling civilizations. What would be a good estimate for this factor? One could argue that eventually all such civilizations would eventually want to venture out and explore space. *However, that may be a purely human motive!* But, it is probable that such a civilization would have a need over time for more resources than are available within one star system. It follows that celestial travel is possibly a long-term requirement for the survival of a species. No matter how much we would like to stay at home, we must eventually either venture out into our world for food and resources as we deplete our domestic supplies or someone must bring them to us if our civilization is to endure. It is reasonable to assume that this same scenario exists for many, if not all, other civilizations. Therefore, an infrastructure must be developed for such survival. It is possible that some civilizations would reach such a point of advancement that any needs for survival could be manufactured right out of the very fabric of spacetime, but such advanced cultures would probably have had to travel in their past to have survived to reach such a point. Hence, all advanced civilizations must reach a point where traveling is required; therefore,

$$f_t = 1. \tag{1.12}$$

Let us assume that civilizations older than one billion years have evolved into godlike or "Q" like (from *Star Trek: The Next Generation*™) and will not be considered in this calculation. From the calculations conducted thus far, about half a billion (see Figure 1.1) traveling civilizations of interest are possibly present in our galaxy now.

Now we need a velocity factor that describes how fast these advanced aliens can travel. Assume the time period of the human civilization to be about ten thousand years (some would debate that three to five is more correct but there are cities dating back even further than ten thousand years). The Milky Way galaxy is about one hundred thousand light years in diameter. If the aliens can travel just one percent of that distance in ten thousand years then the velocity factor for their civilization becomes

$$v = \frac{(0.01)(100{,}000\,\text{lightyears})}{10{,}000\,\text{years}} = 0.1\ \text{light years / year}. \tag{1.13}$$

This velocity is nonrelativistic and is reachable using physics, as we currently know it. In fact, it is possible that spacecraft propulsion technologies currently being studied, such as Solar and Laser Sails, could offer such velocities. The human race will have developed such technology within the next few years to the next few decades. This nonrelativistic speed is much too conservative an estimate. A billion-year-old civilization has a high probability of being much further advanced in the development of propulsion technologies. Even civilizations at least a few hundred years older than our civilization should have much faster propulsion technologies. We will assume this very conservative velocity for now.

Now consider that the galaxy is a two dimensional circular surface area whereas the stars of the galaxy are uniformly distributed about that circle. This is not actually the case but it is a fair approximation of the galactic structure. Therefore, if there are a half billion travelers of interest in the galaxy then the density of travelers, σ_t, in the Milky Way is

$$\sigma_t = \frac{500x10^6 \text{ travelers}}{\pi R^2_{galaxy}} = 0.064 \frac{\text{travelers}}{\text{lyr}^2} \quad . \tag{1.14}$$

Figure 1.2. Theoretical number of traveling civilizations imposed on the galaxy (the spiral galaxy image comes from the Space Telescope Science Institute Public images WWW site)

Figure 1.2 shows the theoretical number of traveling civilizations as imposed on a spiral galaxy typical of the Milky Way Galaxy.

Assuming that the aliens are velocity limited as discussed previously, the number of star systems within the range that traveling alien civilization could visit within ten thousand years is then

$$n_{range} = \pi R^2_{travelers}\sigma_t = 0.064\frac{\text{travelers}}{\text{lyr}^2}(\pi)(1{,}000\text{ lyr})^2 \approx 201{,}062\ . \qquad (1.15)$$

In other words, each traveling civilization could visit at least one of over two hundred thousand distant star systems within each ten thousand year increment. Of course, this assumes that the travelers go from their home world directly to the outer most range. It is very likely that they will pass other star systems on their way. Also, we could assume that the travelers were smart enough to investigate all two hundred thousand systems with their super alien telescopes and that they know which star systems to go to. This seems to be an obvious approach for any star faring aliens. Find a system within range that hosts a civilization then travel to it.

A different way to look at this is as follows. There is a particular density of travelers per light year as given before. Also, these travelers are velocity limited as discussed previously. Assuming the Earth is at the center of a circle that extends to the edge of the velocity limited range of the travelers, then the number of travelers within range of Earth is

$$N_{travelersnearEarth} = \sigma_t\pi R^2_{fromEarth} = 0.064\frac{\text{travelers}}{\text{lyr}^2}(\pi)(1{,}000\text{ lyr})^2 \approx 201{,}062\,. \qquad (1.16)$$

So, more than two hundred thousand traveling civilizations should be in range of Earth! Assuming that they decide to travel at uniformly random times, then the frequency of travelers visiting Earth is

$$F_{EarthVisits} = \frac{N_{travelersnearEarth}}{\text{Travel Time}} \approx \frac{201{,}062}{10{,}000\text{yrs}} \approx 20\frac{\text{visits}}{\text{yr}}\,! \qquad (1.17)$$

Twenty visits per year seems much too fantastic to be readily believed. Of course, the galaxy has a thickness and the density is not uniform. The differences due to those assumptions will change the numbers but by less than two orders of magnitude, which is still two visits every ten years. Using the low end of the error in the approximation would only decrease the frequency of visits to two visits every one hundred years. The calculation suggests very fantastic ramifications indeed! ***This result suggests a high probability of at least one ET visit during the average American's lifetime!***

We could discuss factors that would reduce the frequency of visits. Those factors would include those that suggest technical malfunctions with the traveler's spacecraft, decisions to visit other civilizations that may be in their range as well, we might be too disinteresting for them, as well as many others. However, these factors would all be percentages most likely greater than one percent and therefore would collectively affect the frequency by significantly less than an order of magnitude or so. That still suggests at least one visit within the ten thousand years of the human civilizations existence as a significant possibility. The absolute worst case, assuming none of the factors are zero, is that we are only visited a few times during the lifetime of our star. Also, remember that most of the numbers used here were very conservative. We even considered that the aliens could not travel more than approximately ten percent the speed of light; even we would hesitate to undertake intergalactic travel with such a limitation. Billion-year-old civilizations could have developed to a technology status that would allow us them to visit many more star systems than we assumed for this calculation. The results of the above calculations are depicted in Figure 1.3 and broken down into a bar graph format.

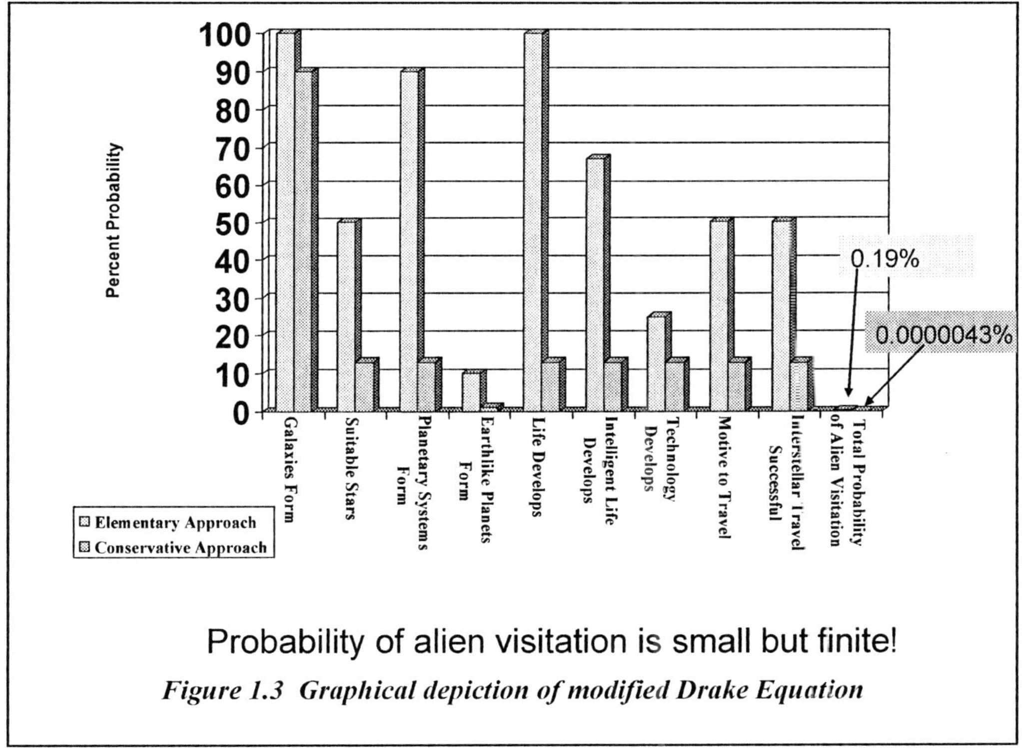

Probability of alien visitation is small but finite!

Figure 1.3 Graphical depiction of modified Drake Equation

The statistical analysis we have just performed indicates a reasonable probability that ET does indeed exist. This analysis leads us to a number of questions that must be addressed. Among those are:

- Does ET exist?
- Is he coming?
- If he does come, what are his intentions?
- Has he already come?

- Why prepare now?

We will address each of those questions one at a time.

Does ET exist? We do not know. We believe there is a high probability that other intelligent life exists in the universe. Anyone who can tell you with certainty that ET exists has access to data and information which we do not have. On the other hand, we do not believe that is possible that anyone can tell you with certainty that ET does not exist. There are too many celestial bodies within the universe about which we have no knowledge. Our technology is far too limited to evaluate the universe for us to make such a flat statement as "ET does not exist"! We do not believe is possible to rule out the possibility of the existence ET with our limited technology. We will have to make great technical advances to answer the question of ET's existence unless he chooses to pay us a visit and make sure we know about it. Let's hope we are prepared if that happens; especially if it is the wrong type of ET.

Is he coming? Once again we do not know. We will attempt to give some reasons why ET might come to earth. We will also attempt to give some reasons why ET would not come to Earth. This too is a case where if someone can tell you with certainty that ET is coming, that person has information which we do not. This is another example of a case where we believe that is impossible for anyone tell you with certainty that ET will not come.

If he does come, what will his intentions be? His intentions will be the result of the type of visitor that he is. We categorize potential extra-terrestrial visitors into four groups.

Those are:

1. Benevolent
2. Neutral or Spectator
3. Researcher
4. Hostile

This is the order in which we would wish to encounter the ET categories. We will examine the potential motivators of each as well as how we should engage each in a later chapter. We will devote a section to our views of how to deal with each of these categories of ETs.

The first of those, the benevolent visitor, is certainly the one we would like to see show up at our doorstep. The benevolent visitor will ideally bring us technical and medical advances beyond our wildest dreams.

The second category is the neutral ET. The neutral ET will come and observe us and interact but not interfere with the progression of our civilization. The presence of a neutral would be quite welcome. What a research opportunity that would provide us.

The researcher will come to size us up so that he can determine whether he can be benevolent, neutral or hostile. That decision will probably be based upon our preparedness primarily in a military sense. He may, however, be more dangerous as he will lull us into letting down our guard thus making us vulnerable to a potentially weaker intruder who morphs into a hostile ET. The best we should expect of a researcher is that he would become a neutral ET. If he were going to become a benevolent ET, he probably would show up as such or at least show that tendency right away. He would only be a benevolent disguised as a neutral to see if we merit his help as a civilization.

The last group is the hostile visitor. This is the guy we hope will never show up; however, there is as high a probability that the hostile ET will show up as any of the other three categories. We had better be prepared for the hostile ET. If we not prepared, the Earth's civilization will be in great jeopardy!

Has he already come? We don't know. There are certainly those among us who say the answer is yes. There is evidence which may be viewed as supportive of their case. There have certainly been many reports of ET spacecraft sightings and of abductions by ET. Obviously some of those events are more credible and believable than others. We have no method for ruling out previous visits by ET.

Why prepare now? Especially, why prepare when we don't even know that ET exists? We should prepare now in order to survive later. For one reason it is possible that he could show up any day. Another is that we have the responsibility to future generations to begin that preparation now because we know that it will be time-consuming. It will be perhaps too late to prepare once ET does show up. It would be a sad day for Earth if a hostile ET were to arrive and we were unprepared. This would also be true if a researcher were to arrive in our atmosphere and quickly decide he could and should be hostile. The only time we have to prepare is now. **Prepare now - survive later.**

1.2 "Eat at Joes" versus "Hide and Watch"

As discussed in section 1.1 it is unlikely that a lesser-advanced civilization could hide from a more-advanced civilization for a significantly long period of time. However, every day the lesser-advanced culture has to develop enables it to develop more capabilities with which to defend itself. This implies that a smart civilization would dig deep into its planet and keep all of its electromagnetic emanations to itself to the maximum extent possible. This suggests that communications systems should be hardwire and underground or advanced beyond our limited broadcast communications system. Only point-to-point open-air transmissions should be made. And absolutely under no circumstances should high power transmissions be sent out omni directionally as we do intentionally on a daily basis.

Assuming that the distribution of the travelers as being friend or foe is uniform, then half of them would do us intentional harm in some way while the other half would intentionally do us good in some way. There is always the potential of even friendly aliens unintentionally introducing harmful organisms into our environment, as was the case with Christian missionaries going to Hawaii. Alien motives would be just that, alien. Most likely we would have no obvious means of knowing what would drive them toward helping or hurting our civilization. We should be very careful before attempting to invite such creatures over for dinner.

This brings us to the mainstream SETI philosophy. For some strange reason the majority of the SETI community believes that a civilization that is advanced enough to conduct interstellar travel would be beyond such things as war or malicious intent. This statement in itself is nonsense due to the reasoning that war and malicious intent may have no bearing on the alien motives. As we build highways to travel from one city to the next, does it bother us tremendously how many insect civilizations we destroy in the process? In our minds, the highway is far more important to our society and culture than the few meaningless anthills that we destroy or displace in the process. These ants are an important part of our ecosystem. The ants might think differently. Most likely, the ants do not understand our motives. Are we more than an insect to an alien? Also, whether we think the aliens are right or wrong does not matter. Right and wrong may be irrelevant topics to the alien thought process. Aliens may have different moral values than we do; our wrong may be their right. The SETI community should understand this, but they continue with the desire to transmit messages out into space with hopes that the "nice" aliens will email back to us the cure for cancer.

What about the "bad" aliens? One should assume that "bad" aliens are at least equally capable of intercepting the transmissions. Should we be constantly flashing a beacon to "Eat at Joe's" as we have been doing since the nineteen thirties? It does not make sound logical sense. Our best strategy for survival in the event there are hostile ET's in transit to Earth or preparing for the trip is to "Hide and Watch" for the travelers while advancing our technology as rapidly as possible.

1.3 Fermi's Paradox: If they are out there, why aren't they here?

In this section we will discuss the so-called Fermi's Paradox and details of various types of population models. First we will discuss the general premise of the paradox and the mathematical basis upon which it was based. We will then demonstrate through more updated models for population growth and species interaction that Fermi's Paradox is not really a paradox at all. The basis for Fermi's premise was a mathematical model that fits well for nuclear decay processes and reactions but very poorly for population growth in a "realistic" universe. We will then discuss more likely scenarios and give models and simulations that will shed more light on the topic.

In the late 1950s the Nobel Prize winning nuclear physicist Enrico Fermi posed a very interesting argument to his lunchtime colleagues. Fermi theorized that if there were intelligent civilizations elsewhere in the galaxy then they should have long since colonized the entire galaxy. Fermi's reasoning was very sound and based on simple population expansion models.

Consider the exponential population equation, also known as the Malthusian Population model,

$$p(t) = p_o e^{kt} . \tag{1.18}$$

Where p(t) is the population of a species as a function of time t, p_o is the initial population, and k is the population expansion coefficient. Assuming that the initial population of the human race was one individual approximately ten million years ago and the present population of the world is about six billion people, the expansion coefficient is found as

$$p(10x10^6) = 6x10^9 = 1e^{k(10x10^6)}. \quad (1.19)$$

Taking the natural log of both sides and solving for k yields

$$\begin{aligned} \ln(6x10^9) &= k(10x10^6) \\ k &= \frac{\ln(6x10^9)}{10x10^6} = 2.25x10^{-6} \end{aligned} \quad (1.20)$$

Therefore, the population growth model for the human race is approximately

$$p(t) = e^{(2.25x10^{-6})t}. \quad (1.21)$$

Using the population model in equation 1.21 we can determine the approximate length of time it will take us to overpopulate Earth and have needs to expand to new planets or become extinct due to the consumption of our critical natural resources. Earth has a surface area of about $5.1x10^8$ km^2 and roughly two thirds of that is water. So, the Earth's useable surface area is about $1.7x10^8$ km^2. One of the world's most populated countries is Japan, which has a population density of about three hundred fifty humans per square kilometer. For this calculation we will assume that we could handle five hundred humans per square kilometer before overpopulation drove us all mad. So, the maximum population Earth could support is

$$p_{max} = 500\frac{\text{humans}}{\text{km}^2}\left(1.7x10^8\,\text{km}^2\right) = 8.5x10^{10} \approx 1x10^{11}\,\text{humans}. \quad (1.22)$$

Figure 1.4 shows a log-log plot of the human population growth. The maximum human population for Earth is about $1x10^{11}$ humans. The figure shows that the human race will reach this maximum population in just a little more than ten million years. Recall that the galaxy is about ten billion years old.

Assuming that all planets habitable in the Milky Way Galaxy are Earthlike, then we can determine that the rate of expansion of our species is one planet per ten million years. Of course, this assumes that there was an initial population and no delay for life sparking is assumed. The Earth was here for five billion years or so before humans walked on it. However, now that humans are here, we see that from Figure 1.4 that in just a million years or so at our present population growth rate we will have completely overpopulated Earth and will need to reach out to other celestial stations to sustain ourselves. The alternative will be a retrenchment in technology and probably population size accompanied by reduction in life expectancy.

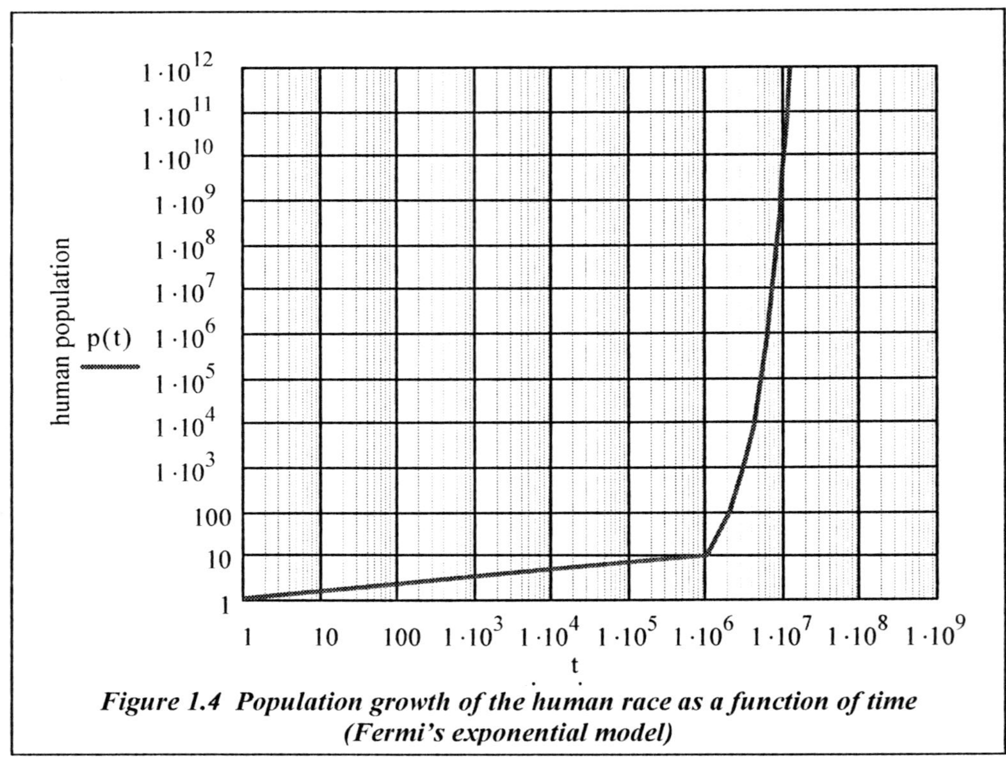

Figure 1.4 Population growth of the human race as a function of time (Fermi's exponential model)

The slope of the curve will vary locally with time as a result of wars, plagues, famines, political actions, etc. The technology needed for such galactic travel will probably be developed long before it becomes critical. Of course, there must be sufficient planning to allow roundtrip travel in time to set up the required supply routes. It is reasonable to assume that we would send unmanned exploration vehicles before attempting to send out humans if we develop technology soon enough.

Earthlings will enter the new Marco Polo era. Members of the civilization may be driven by greed long before need. There is always the possibility that we will choose to acquire resources from uninhabited celestial bodies as opposed to approaching those with civilizations. However, if food is our objective for becoming aliens, we will probably be looking for less technologically advanced and less powerful civilizations to exploit, as long-range farming would prove difficult to sustain. We might look for uninhabited worlds with environments favorable to agriculture as we understand it. However, we should expect that planets with an environment favorable to our agricultural techniques will probably be inhabited by intelligent lifeforms already. What kind of aliens will Earthlings be? The answer to that question will probably be driven by our motives for travel.

Figure 1.5 shows a graph of the human race's expansion as the number of planets populated to maximum density versus time in years. The figure shows that provided the human race continues to expand at its present rate, we will have populated over four hundred billion planets in just about twenty three million more years. If every star system in the Milky Way has a habitable planet, it should only take us twenty three million years from now to have colonized them all! Again, recall that the galaxy is only about ten or so billion years old.

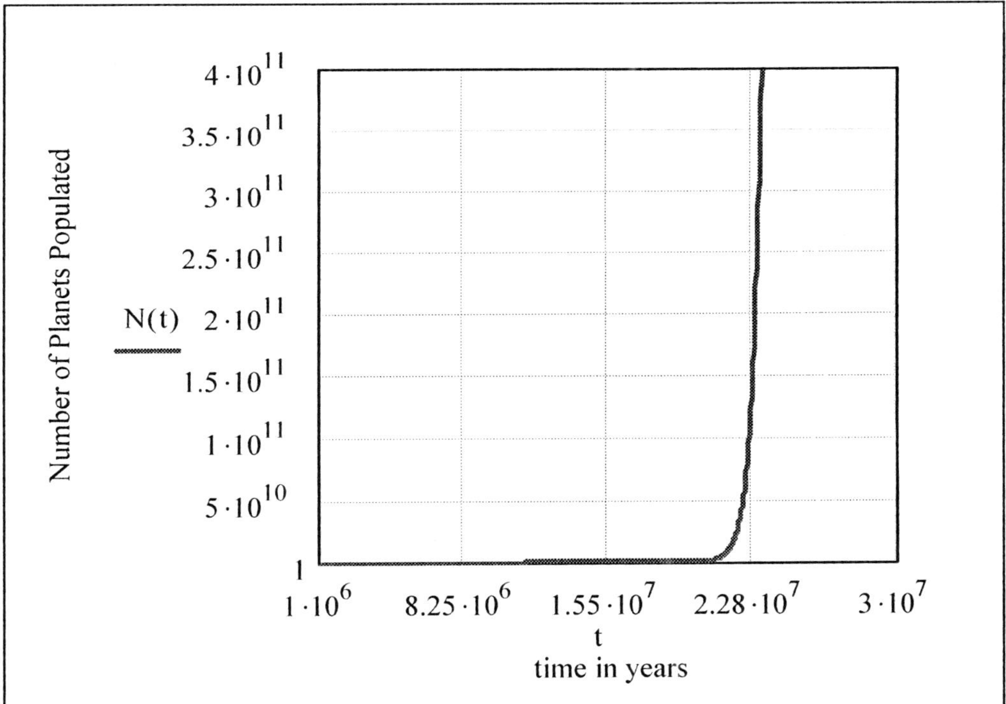

Figure 1.5 Number of planets populated by the human race as a function of time (Fermi's exponential model)

Now Fermi's Paradox becomes apparent. Why have we not been colonized by a species geologically only a few tens of million years older than ourselves? Through this reasoning Fermi concluded that we must be alone in the Universe.

So, was Fermi's conclusion correct? Many people from various walks of life and technical professions have debated on both sides of the argument that Fermi was right or wrong. Fermi's conclusion being correct is the simplest solution, albeit a rather perplexing one. Zero is quite often one of the many and one of the easiest solutions to mathematical problems. The other solutions are usually slightly more difficult to obtain. We shall now consider other possible solutions to Fermi's Paradox.

The SETI community has produced a myriad of papers about Fermi's Paradox. Some of the solutions that have been developed are things such as:

- a Prime Directive prohibiting interference with primitive cultures,
- lack of interest in us,
- no previous civilization has had a need or desire to colonize the galaxy,
- we are the first civilization to develop,
- we are already the results of a colonization effort,
- it is too technologically difficult,
- we have been overlooked,
- or perhaps we are being visited on a regular basis and just do not know it.

Any or all of these suggestions appear to be viable solutions to the Paradox posed by Fermi. However, there is no overwhelming evidence to support any of them.

What might keep us from expanding and colonizing the galaxy at such an exponential rate? Interest and difficulty are two possibilities. Consider the naval explorers who crossed the oceans. Only a few well-funded and daring explorers made the trips at first. Others followed them many years later as the ship technology improved and decreased the risk significantly. Of course, these colonists had some reason for leaving the comforts of home on the European continent. Usually, political or religious pressure or dreams of wealth drove them to leave home for the unknown. The unknown promised freedom, land, riches, etc. The settlers of the "Old West" were driven in a similar manner. The difference between these daring colonists and spacefaring colonists is a matter of technology. The technology required to colonize a planet once you are already on that planet is fairly simple. On the other hand, to get from one planet to another, and more importantly, one star system to another is quite another matter indeed.

It is possible that we could reach the goal of inventing a warp drive and building a starship within the next fifty to one hundred years provided that sufficient planetary resources are focused on that goal. But do we as a civilization have the desire to venture out to the unknown and attempt the colonization of other worlds? Once the technology for building starships is as commonplace as that for building airplanes, it is likely that some groups will dare to attempt galactic travel. "Build it and they will come." From the previous calculations, it was shown that we have a few million years before we reach overpopulation. We will begin depleting our critical resources and taxing our food production capabilities long before that. We are already rapidly depleting our fossil fuel energy sources; they will become scarce in a few years, meaning that we need to develop or find alternatives fuel sources for at least some applications. Hence, it is likely that we will perfect and make economical starships available to the public in that timeframe. So, if it is reasonably possible that we will go in just a few short million years, then what about those civilizations older than us?

It would seem that a person could go insane trying to solve Fermi's Paradox. There is no known theoretical or modeling approach that can really offer insight into its solution using only Fermi's model. Perhaps this is why it is appropriately called a paradox. But, there must be a solution out there. The way to solve the problem is by finding experimental evidence that gives an indisputable resolution to the paradox. Find evidence of a technological civilization out there, then we will at least know that the zero solution is incorrect. At that point we should be prepared for any of the best and worst case solutions.

What if we were the second civilization to develop and we are only a few million years behind the first? We could soon be faced with the onslaught of the first wave of colonists from that civilization. What if many similarly aged civilizations are presently fighting each other to hold off their respective colonization waves? Perhaps one day we will be sounding the charge of the first Milky Way colonization wave. We have no idea what the case is at this point. The possibilities are endless until we have further evidence.

Fermi's Blunder

It should also be discussed here that Fermi's Paradox assumes an exponential growth of one species. There are at least two major flaws with this initial assumption. The first is that no species ever discovered on Earth follows a simple exponential growth model. Secondly, it is very likely that the galaxy is teeming with many species that maintain their own respective niches as well as many that are in competition over certain niches.

Nature here on Earth offers many examples where the struggle for existence between two similar species fighting over the same niche (food supply, space, etc.) occurs. Ultimately one species wins out by causing the complete extinction of the other species. This phenomenon is known as the "principle of competitive exclusion" and was proposed by Darwin in 1859 in his *Origin of Species*.

There are also cases on Earth when the "principle of competitive exclusion" is in direct contradiction with some well-known natural phenomena. An example of one of these natural contradictions is called the "plankton paradox" and is focused on the variability of plankton organisms which all seem to occupy the same niche. All plankton algae use the same niche, which consists of solar energy and minerals dissolved in their native habitat waters. There are many plankton algae species, many more than the different types of mineral components in the water habitat of the plankton.

Therefore, it is commonplace that multiple species compete over the same niche and multiple species survive indefinitely. Also, there are predator-prey competition examples here on Earth where multiple predators attack multiple prey and the prey species survive indefinitely. These complex population systems are dynamical systems with many variables and forcing functions. The number of possible solutions is nearly endless depending on the initial conditions and the unexpected imposed conditions of the systems. It appears they will peacefully coexist as long as there is sufficient food for all of them. What will happen if the food supply begins to be limited? Do we then see another demonstration of survival of the fittest on a broad basis?

Consider the following system of differential equations

$$\begin{aligned} y' &= ay - bz \\ z' &= cy - dz \end{aligned} \quad (1.23)$$

This is a simple model of a competing predator-prey population system. The value y represents the predator population and z the prey population, a, b, c, and d are constants based on the particular niche. Depending on the initial conditions and the values of the constants

$$\begin{aligned} y &= e^{pt}\left[y_o \cos(qt) + \frac{y_o}{q}\sin(qt)\right] \\ z &= \frac{a}{b}e^{pt}\left[z_o \cos(qt) + \left(\frac{ay_o}{b} - z_o\right)\sin(qt)\right] \end{aligned} \quad (1.24)$$

is a specific set of solutions to the system of differential equations in equation 1.23. Here p and q are constants arising from the solution process and y_o and z_o are the predator and prey initial populations. Again, equations 1.24 are a specific set of solutions, there are many solutions that can be arrived at numerically. However, using equations 1.24 in a simulation will suffice to make the point that sometimes the preys survive and sometimes they do not.

Figures 1.6 and 1.7 demonstrate that the dynamical system is sensitive to the initial conditions and various other variables describing it. Therefore, the simple Malthusian or exponential population growth as described previously is a drastic oversimplification. Perhaps Fermi's Paradox is not as paradoxical as it initially seems. One could imagine that the galaxy is much like Earth with multiple species supporting and competing against each other over various niche resources. Perhaps the society that is a few million years older than us is not preying on us as often as expected because they are defending themselves from predators a few million years older than them. The possibilities are limitless. Let's hope that we are living in a natural environment, as on Earth, where coexistence of predator, prey, and other competing species is possible. Also note here that in essence the solutions to these equations become the Lotka-Volterra inter-specific competition equations (feel free to look it up as an exercise for the reader).

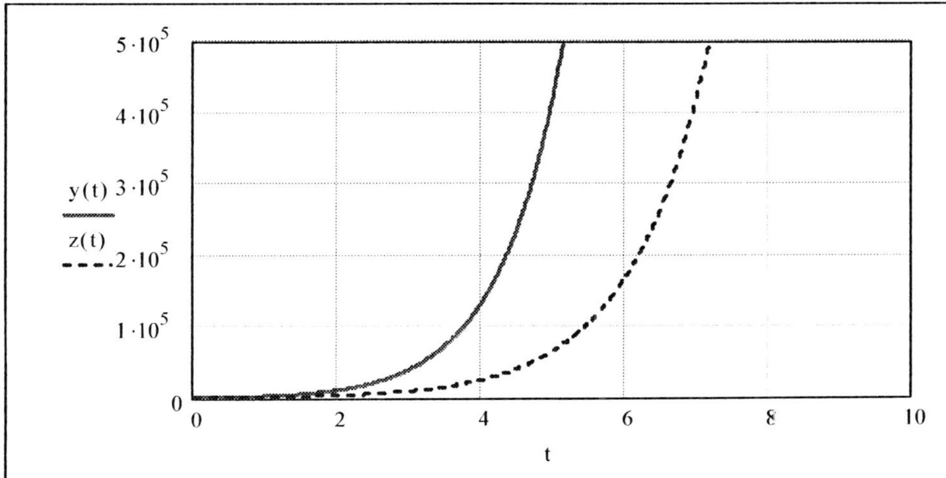

t is in arbitrary time units

Figure 1.6 A situation where both predator and prey thrive (Note that the predator and prey initial populations are equal)

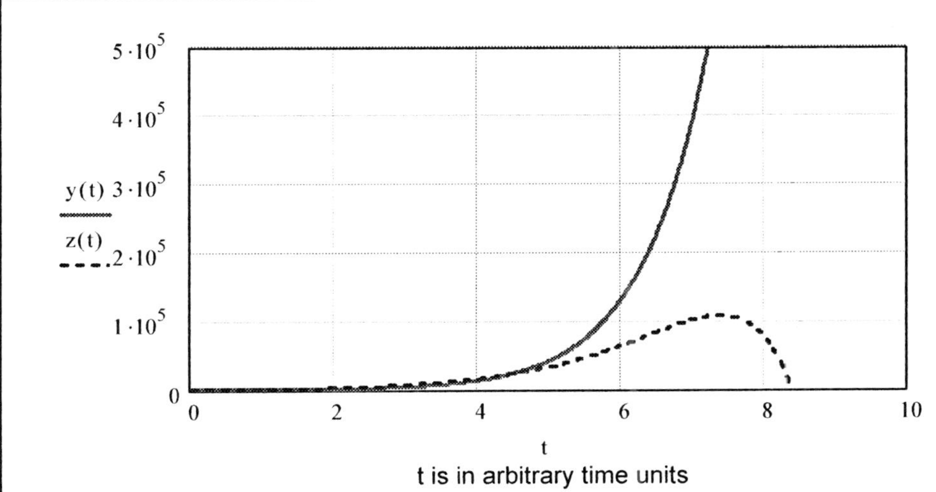

Figure 1.7 A situation where the predator destroys the prey population (Note that the predator population is initially 10 times larger than the prey population)

The Baker's Yeast Example

We have discussed the simple exponential growth rate population model that Fermi assumed as well as a slightly more complex one with two competing species. To illustrate a point here about a single species we will give a simple experimental example of yeast species growth and how it grows as a function of time.

Baker's Yeast or *Saccharomyces cerevisiae* is a microscopic, single-cell organism which is classified as a fungus. Yeast exists on plants, in the air, in soil, and in and on humans and animals. Yeast is the living substance responsible for the production of the enzymes that permit fermentation. The action of yeast is to metabolize simple sugars and to convert them into alcohol with heat and carbon dioxide as products. The carbon dioxide bubbles get trapped in the gluten strands of bread, causing it to rise. Yeast is commonly used in useful processes, such as brewing and dough rising. It is used to produce vinegar as well as in bread, wine and beer making (major staples of any civilized culture).

By taking a small packet of Baker's Yeast that can be found in any supermarket and activating it with warm water and sugar, an illustration of why Fermi's Paradox does not follow the real world becomes apparent. Figure 1.8 shows pictures taken through a microscope at different times following the Baker's Yeast activation. As the yeast colony grows the pictures show the increasing population. Note that the tiny spherical objects in the images are the individual yeast.

Figure 1.8 Images of a Baker's Yeast colony's growth

Figure 1.9 shows the yeast population growth as a function of time. The important point to note here is that the growth curve does not follow a simple exponential population growth model. Instead, the data plateaus after a period of time. This is not surprising actually. This plateau is common in all species known to man and in mathematical terms growth follows the Logistics model.

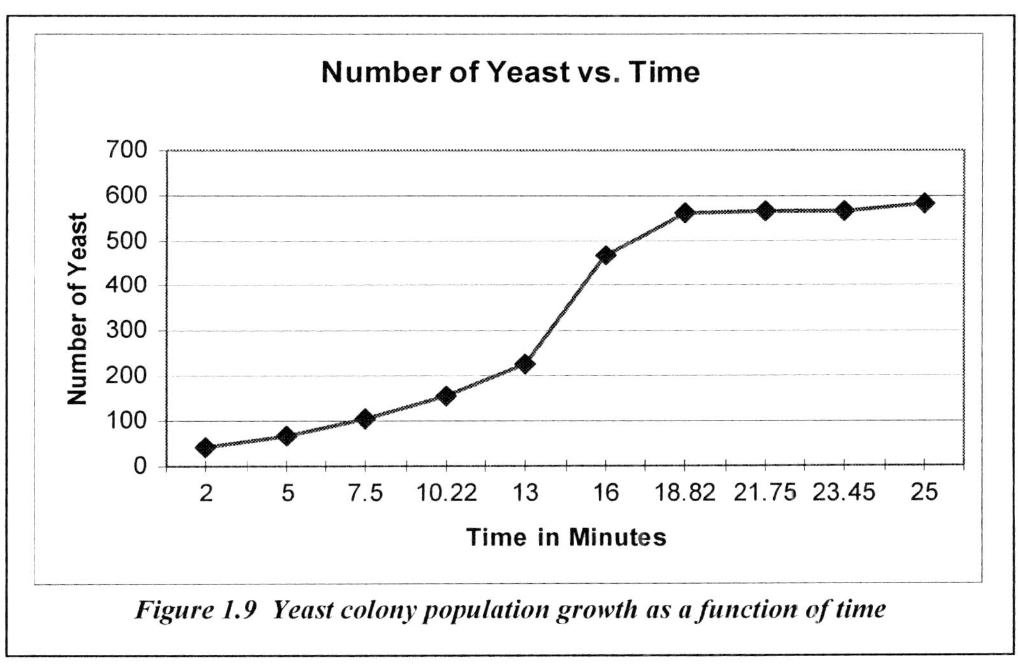

Figure 1.9 Yeast colony population growth as a function of time

Equation 1.25 is the Logistics model for population growth, first published by P.F. Verhulst in 1838. The population as a function of time, t, is

$$P(t) = P_o K / \{P_o + (K - P_o)\, e^{-rt}\} \qquad (1.25)$$

where P_o is the initial population, K is the do not exceed population or maximum population, and r is the growth rate coefficient. Note that a curve fit to this data gives values for P_o of 26.648, K of 550, and r of 0.4472.

The point of putting this simple experiment in this section is to illustrate how populations of species grow in the real world. If the species were observed for longer periods of time and there were no outside intervention of adding more food and energy into the environment then the yeast would begin to die off and exponentially decay back towards zero and extinction. Also note that there was no other species involved and therefore no predator-prey or niche competition involved. So the conclusion that can be gathered from this simple experiment is that Fermi's Paradox is not a paradox at all. The theory is simply flawed from the most basic assumptions made by Fermi in the onset of the discussion.

Another population model that is well known within the military community is the so-called Lanchester's laws. In 1914 Fredrick Lanchester developed laws for modeling combat based on population growth. The Lanchester models were a set of coupled differential equations much like the Lotka-Volterra equations with the added advantage of allowing for multiple sets of species with different strength levels. These models were developed in great detail to discuss force-on-force combat simulation. The models have been improved over time and will be discussed in later sections of this text.

The goal of this discussion was to make obvious Fermi's Blunder, which should be apparent to the reader at this point. Various detailed and developed models of populations and inter-specific competition were available to Fermi in the 1950s, yet he chose to use a mathematical model that had been shown to be a "poor" description of real world populations. Therefore, discussion of Fermi's Paradox should immediately begin with a discussion of Fermi's Blunder. Realization of the blunder quite readily deflates the paradox.

Aspects of Humanity's Population upon ET Interaction

We have discussed population growth from the simple population model to slightly more complex competition models. Now we will discuss the various components within a given population model that may affect it, including the possibility of an alien interaction.

An important objective of any intelligent society should be to sustain its population. That society should also have an objective to grow within reasonable limits. In order to achieve those objectives, it is important to understand the factors that influence the population's dynamics.

We must understand the classical contributors to population trends before we can address the mechanisms by which a visit by ET could influence the objectives of sustaining and growing our society's population. To that end we have developed a population model to examine the contributors. This model recognizes the existence and importance of both continuous (enduring) and transient (short-term) factors. The continuous functions which have a persistent statistical basis are the birth rate and the mortality rate over time. The transient functions in our model are 1) wars, 2) energy sources, 3) food supply, 4) medical breakthroughs and 5) disease. Each of these functions will be addressed independently.

The birth rate seems intuitively obvious. It is the number of births per member of the populace per year. The birth rate is the engine which drives the population to sustain growth. If the population is 1.0, the birth rate must be greater than 1.0 plus the mortality rate for the population to grow. The desirable range of birth rates is calculated as

$$\text{Birth Rate} > 1.0 + \text{Mortality Rate}$$

A smaller birth rate leads to a stable or worse yet a diminishing population.

The mortality rate is the sum of all continuous causes of death. In this model the mortality rate includes death from such factors as old age, accidents, obesity, drugs, alcohol and murder as well as routine maladies such as heart disease and cancer.

Our model includes and recognizes the importance of major transient functions. A major transient function must last long enough to attenuate the continuous functions over a significant period of time. We have grouped the transient functions under the categories of:

- War
- Energy Sources
- Food Supply
- Disease
- Medical Breakthroughs.

These factors in general with the notable exception of medical breakthroughs tend to be additive to the mortality rate. They may also carry a resultant side effect that manifests itself as a reduction in the birth rate during their span of influence.

War obviously decreases the birth rate while it is being waged as many men are away from home for a prolonged period of time. War has also been shown to generate a pent up rebound and correction in birth rate at its conclusion when the men return home. The birth rate typically surpasses its prewar level as a large spike for a short period after the men get back home.

The level of the food supply limits our ability as a global society to feed our population. The factors that dominate the food supply are famine and agricultural advances.

Famine is the result of long-term reductions in crop development and maturation. Famine may result from sustained meteorological events such as El Nino. Famine might also result from abnormally high levels of crop predators. The presence of famine adds to the mortality rate while at the same time reducing the birth rate due to the debilitation of starvation.

Agricultural advances allow us to produce more food per unit of land under cultivation. Agricultural advances tend to decrease the mortality rate. Agricultural advances have over time had the effect of reducing the birth rate because less children were required to produce the crops and because there was more time for out-of-the-home recreation.

One might argue that distribution methodologies are also a limiting factor. It is true that until the last half century or so that we as a society were hampered by distribution system issues in an effort to feed some remote tribes within our society. Abundant fossil fuel sources have provided us the ability to reach even the most remote tribes with nutritious food stuffs. This will continue to be true as long as our fossil fuel stores last or if we develop or harness reasonable substitute or alternative fuels to drive our machines. New sources will be necessary because we are consuming fossil fuels at a rate far greater than that at which they are being produced.

Disease is defined as the incursion of a new widespread, highly contagious pathogen into the society. That pathogen must be a deadly agent for a number of years or spread with incredible rapidity through the society in order to significantly increase the mortality rate. Such an explosive increase in mortality rate would be expected to carry an attendant decrease in the birth rate due to the rampant decease in the number of adults who would have otherwise been potential parents. Another mechanism for reducing the birth rate would be the introduction of a highly contagious pathogen which causes mass sterility. A huge increase in sterility will cause meaningful decrease in the birth rate for many years even if it persists for only a few years as it would probably affect members of the society of all ages. The society would have to age for a period on the order of thirty years to see the birth rate rebound. It is possible that a single pathogen could cause a simultaneous increase in mortality rate and decrease in birth rate.

Medical breakthroughs can and do eliminate or significantly reduce the influence of a source of death. They may also increase the general fertility of the society. Increases in fertility can be expected to increase the birth rate. Medical breakthroughs are the one transient function that can often decrease the mortality rate. They have the effect of increasing the growth of the population.

A visit by ET, should it happen, would possibly produce some effects on the global population. In some cases the effects will be driven by the intentions of the ET. The effects may not always be what were initially intended.

A hostile ET would almost certainly cause an increase in the mortality rate through acts of war if he chose to attack us directly. Instead of a direct frontal attack he might introduce virulent new pathogens into our society to weaken our defensive capabilities. He might have the ability to influence our weather cycles to produce famine. He might have the ability to disrupt the flow of fossil fuels. All of those actions would increase the mortality rate and restrict the growth of our society's population.

A benevolent ET could bring medical breakthroughs that would significantly decrease the mortality rate to have a prolonged increase the population of the global society. He might give us the secret of a fuel source to serve as an alternate to fossil fuels. These actions would have a net positive effect on the growth of the population through decreased mortality rate and perhaps increased birth rate.

A visit by any type of ET could bring one or more foreign pathogens. Although planetary biologists continue to debate the possibility, it is conceivable that those pathogens could affect our society in the same manner as those carried by the well intended missionaries to the Sandwich Islands. The bacteria carried by those missionaries almost wiped out the population of the islands. Figure 1.10 gives a fairly complex model of population growth due to positive and negative impacts including alien interaction. The alien interaction might be an aspect in the war filter, or the disease, or medical advances, who knows what type and how alien interaction will impact the population system. This model is a standard "linear systems" model used in most engineering fields to describe complicated multivariable systems of systems. If other variables arise that are not shown on the diagram it is simple enough to add another linear combination or loop with a box for that variable. Also, each box may consist of a complex model as

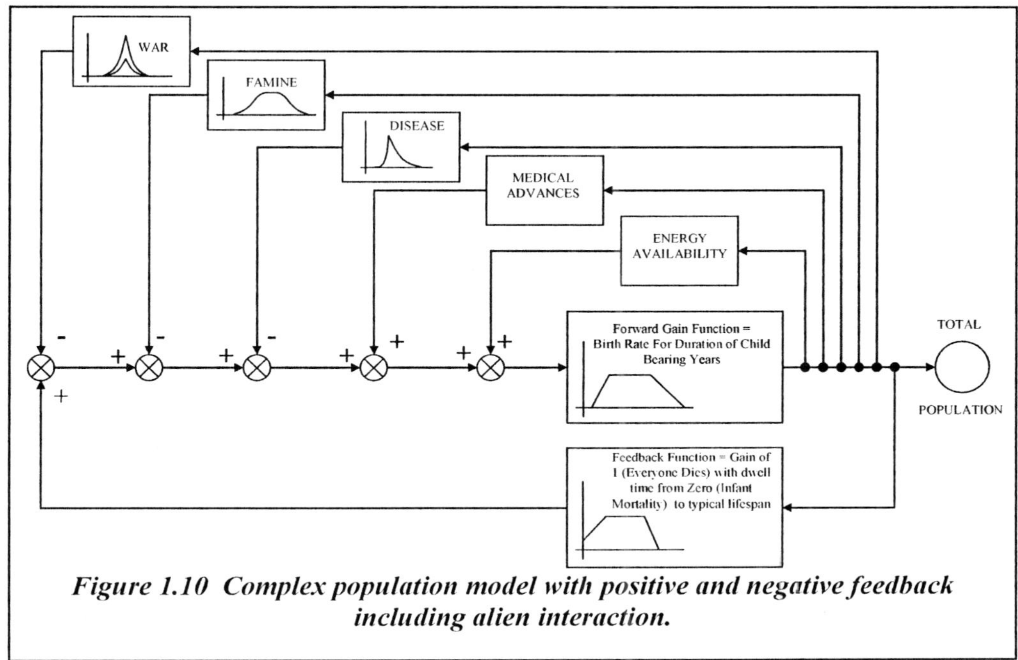

Figure 1.10 Complex population model with positive and negative feedback including alien interaction.

complicated as the one shown for the complete system in Figure 1.10. This is a very good approach for describing the interaction of such an impact as alien attack. More efforts should be spent developing this model in greater levels of detail.

1.4 Distance Barriers and Power Requirements for Starfaring Aliens

For now we will assume that for some reason we have yet to be part of a full force colonization/attack wave from an extra-terrestrial civilization. It is possible that there have been travelers who visited us and perhaps even meddlers that have involved themselves in our history. It is difficult to prepare for an enemy whose existence is unconfirmed and whose strengths and weaknesses are unknown. We will try to make some educated guesses as to the types of capabilities an alien attack force might have. .

To make the discussion simpler we must first categorize our possible attackers. We will categorize the alien attackers by their ability to manipulate energy. How much energy their civilization is able to mobilize to the front lines of an attack wave is how they will be categorized. A scale based on Power already exists and was developed by the Russian Astronomer Nicolai Kardashev in 1964.

Kardashev described five types of civilizations based on the power in watts that they could manipulate. The types are listed as Roman numerals I, II, III, IV, and V and were assigned as follows with equivalent popular science fiction examples:

Type I:	10^{16} to 10^{26} watts	Federation of Planets from *Star Trek*
Type II:	10^{26} to 10^{36} watts	the Republic/Galactic Empire of *Star Wars*
Type III:	10^{36} to 10^{46} watts	the Ancients from *Stargate SG-1*
Type IV:	10^{46} to 10^{56} watts	the Q from *Star Trek: The Next Generation*
Type V:	10^{56} to 10^{66} watts	the Timelords of *Dr. Who.*

Carl Sagan later added to Kardashev's Scale by interpolating the values into the logarithmic functional form

$$K = \frac{\log(W) - 6}{10} \tag{1.26}$$

where K is the Kardashev Type and W is the power in watts the civilization can manipulate. Figure 1.11 is a graph of the Kardashev scale.

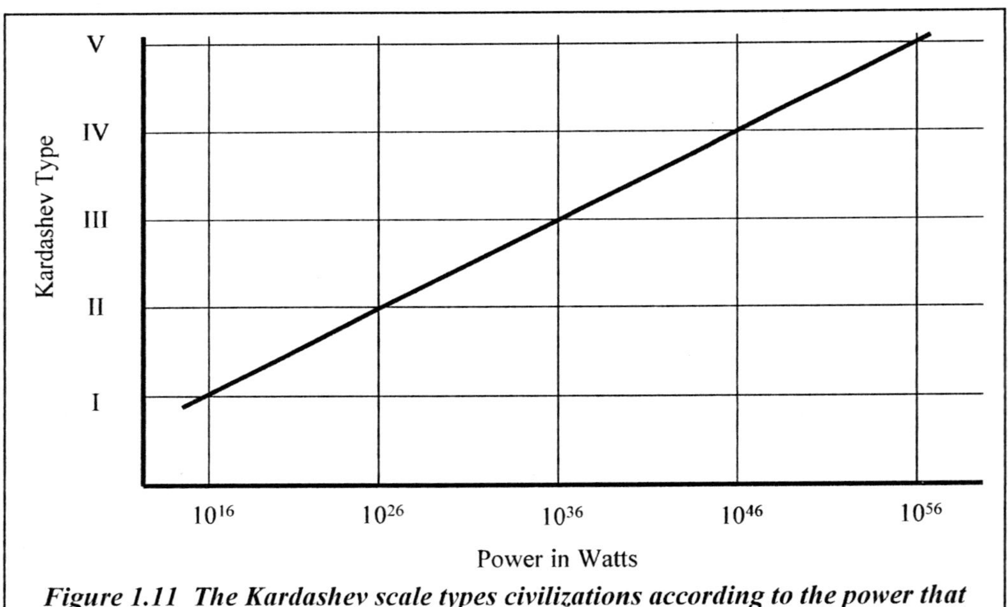

Figure 1.11 The Kardashev scale types civilizations according to the power that civilization can manipulate

There has been a tremendous amount of work defining and attempting to add detail to the Kardashev scale and we will not discuss those efforts here. The problem with the Kardashev scale is that it is limited to levels of power manipulation. This will work well for studying destructive capabilities of a military but more details are needed when it comes to discussing how rapidly that force can be projected. We suggest here an addition to the Kardashev scale that is based on velocity of the civilization's force capabilities. In other words, how fast can they transport their military from point A to point B.

We will use a scale based on velocity since velocity is a function of energy (and therefore power) as far as modern physics understands. Who is to say what energy is required to achieve extremely high velocities millions of times greater than light speed? At present date General Relativity suggests that in order to create faster than light capabilities via a spacetime warp, exotic or negative energy on the order of the mass of a star would be required (some suggest the energy is more like on the scale of a galaxy but this is presently being debated within the physics community). So, for discussion purposes we will simply use a velocity energy equation that does not take into account exotic General Relativistic or even Quantum Mechanical phenomenology. The scale is not meant to be a physical description of nature. Instead, it is simply for description of the alien civilization's capabilities.

Consider the following velocity equation

$$v = \alpha c \tag{1.27}$$

where v is the maximum velocity the civilization can reach, α is the warp factor, and c is the speed of light. Since humans are a subwarp society presently, the warp factor α is significantly less than one. Table 1.1 shows a description of alien species based on their warp factor. For simplicity, subscripts were added to α to represent the civilizations power level. The higher subscripts represent the more powerful civilizations. Table 1.1 also shows that the α_n is an exponential function, which can be used to describe a civilization's power level specifically.

Figure 1.12 shows the α values overlaid on the Kardashev scale. Note that a civilization with the capability to traverse the galaxy from one side to the other in one day falls within the Type III category. Types above Type III are beyond discussion at this point. Also note that the calculation to place the α values on the Kardashev scale assumed that the alien spaceship was 10,000 kg in total mass and could reach the warp velocity for the respective α_n requirements.

Table 1.1. Species Classified by Warp Factor

$\alpha_0 \sim 0.1$	$\alpha_1 \sim 1$	$\alpha_2 \sim 100$	$\alpha_3 \sim 10^3$	$\alpha_4 \sim 10^6$	$\alpha_5 \sim 10^8$
Galaxy transit time >400,000 years	Galaxy transit time ~400,000 years	Galaxy transit time ~4,000 years	Galaxy transit time ~400 years	Galaxy transit time ~1 year	Galaxy transit time ~1 day
Solar System Explorers	One-way Colonists Worldships	Explorers Local Cluster Expansion	Unstoppable Quadrant Colonists	Unstoppable Galactic Colonists	Godlike

Note that α is an exponential function and can be carried to higher power level classifications. Curve fitting the endpoints yields the equation

$$\alpha_n = 0.1e^{(4.1)n}$$

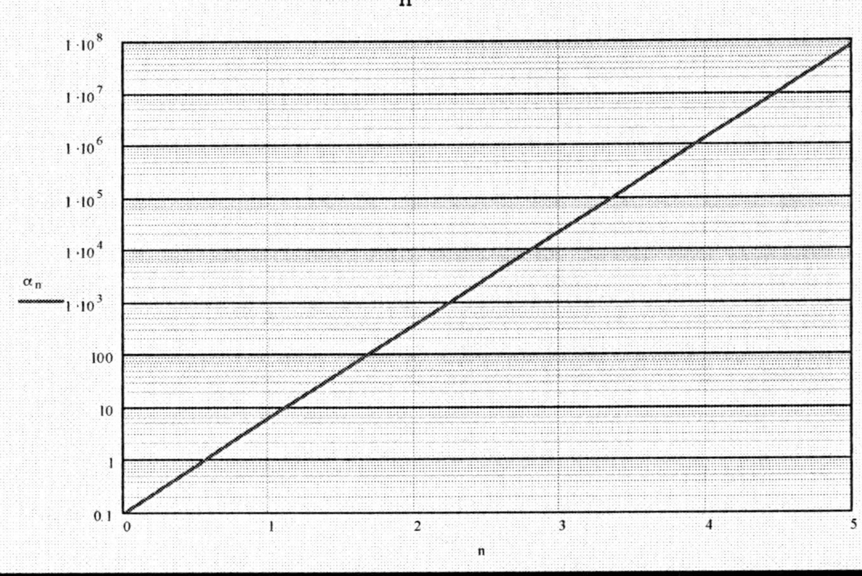

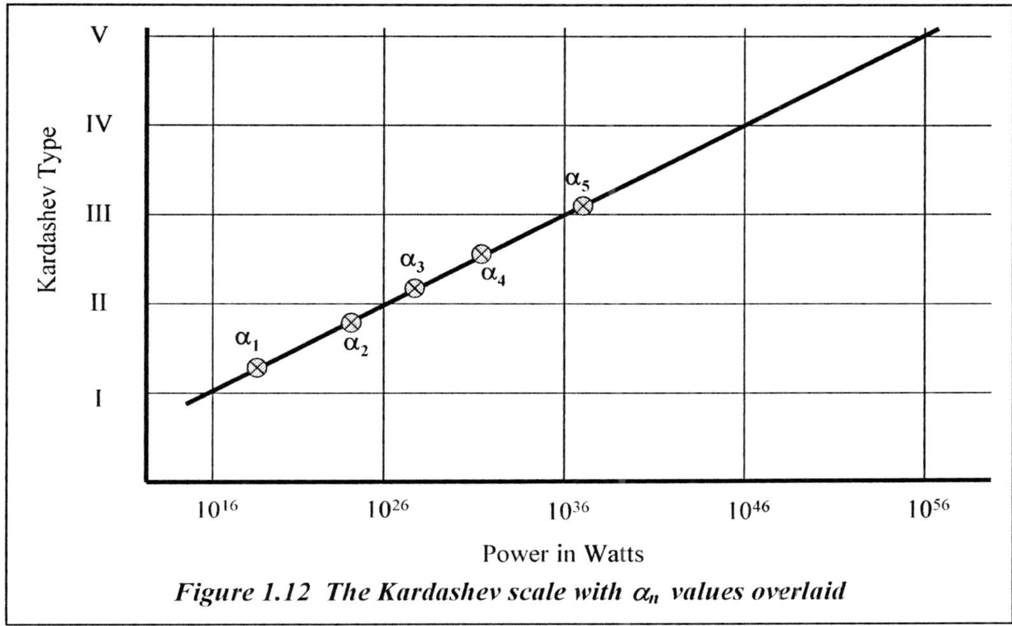

Figure 1.12 The Kardashev scale with α_n values overlaid

Fermi's Blunder Once Again

So let's consider our galaxy only for this discussion. Most people think Fermi's Paradox gives them plenty of reason for saying that we are alone in the universe. We say that this means the numbers haven't been considered and the math completed. Here is why. Consider the five types of aliens described as follows:

α_1 can reach a velocity of 1c
α_2 can reach a velocity of 100c
α_3 can reach a velocity of 1,000c
α_4 can reach a velocity of 1,000,000c
α_5 can reach a velocity of 100,000,000c.

Let's also assume that they have telescopes good enough that would enable them to see any planet in the galaxy with sufficient resolution so that buildings could be discerned. This would signify to the aliens that civilization exists on the planet they are viewing.

The human race did not construct buildings of any real concern (to our knowledge) earlier than about 10,000 years ago. So, let's assume that an alien race sees a planet with buildings on it, then they start for it. In other words, when our first structures went up here on Earth the aliens began traveling here as soon as they viewed this activity.

For the α_1 who can only travel at the speed of light, this means that they could not be further away from Earth than 5,000 light years. It would take that long for the light to leave Earth and reach them and then that long again for them to travel from their vantage point to Earth.

For the α_2 the light from Earth could travel as far as 9,900 light years from Earth to the alien telescope since it would only take them 100 years to travel the distance from their telescope to Earth.

For the α_3 the light from Earth could travel as far as 9,990 light years from Earth to the alien telescope since it would only take them only 10 years to travel the distance from their telescope to Earth.

For the α_4 and α_5 the light from Earth could travel as far as 9,999.9 light years from Earth to the alien telescope since it would only take them less than a year to travel the distance from their telescope to Earth.

The conclusion from this argument is that the aliens even as advanced as the $\alpha_{2\text{-}5}$ must not live within a radius of 10,000 light years of Earth and the α_1 must not live within a radius of 5,000 light years from Earth. Figure 1.13 shows this distance overlaid on the galaxy. Note that it is only a tiny portion of the galaxy.

Oh, but they wouldn't have to be aware of us. They would visit every planet if they were advanced enough. Okay, let's consider that argument for a moment. There are

Figure 1.13 *Portion of galaxy where the "godlike" aliens should be aware of Earth's civilization. Must be no aliens interested in us within this region!*

somewhere between 100 billion and 3 trillion stars in the Milky Way galaxy (astronomers argue this continuously). Let's assume a conservative estimate of 100 billion stars. For the α_5 that can travel anywhere in the galaxy in less than a single Earth day, it would still take them an inordinate amount of time to visit each star in the galaxy. Even if they could visit 100 stars a day due to proximity, then it would still take a billion days to visit every star in the galaxy. That is about 2.7 million years of doing nothing but bouncing from star to star as fast as possible. If these aliens showed up on Earth more than 100,000 years ago they would have missed even the best cave abodes. Had they visited before that they would have had a fun day at the zoo seeing all the neat animals and funny primates and hominids. And, we wouldn't be the wiser of the visit.

Unless they visited within the last 60 or so years we probably would not have been able to detect them unless they actually came down and hung out with us for a while. And why would they do that? To an α_5 we would be nothing more than ants. They might come to study us, remove us, or stumble into our hill, but unlikely they want to come stay with us and hang out – but who understands the mind of ET (all of these things are possible)? It should also be recognized here that the lesser advanced aliens would require billions of years or millions of vessels to visit the galaxy star by star.

Fermi suggested the simple population model overrun of the universe by advanced aliens, but his simple model only considered one species. As discussed previously, more realistic population models with multiple species both competing and noncompeting there would be bubbles within the population space that the advancing species would never get around to just as we haven't overtaken the Congo or the Amazon Jungle or various miles of desert lands or oceans or Yosemite National Park (although the latter is visited quite often but not lived in).

Indeed, when the math is really and truly considered Fermi's Paradox is no paradox at all. In fact, it is surprising to the authors that the paradox has survived peer analysis for so long. We only considered our galaxy here and used mainly "godlike" aliens for our discussion. When allowing for the entire universe and lesser advanced aliens Fermi's so-called Paradox is not even applicable and realistically *makes no sense.*

One final note: If we consider that aliens wait until they detect an electromagnetic signal from us before they head on over for a visit, then this drastically reduces the radius within which there would be no aliens. The radii for the levels of aliens ranges from 35 to 70 light years from Earth. This is a very small circle indeed. Once again, Fermi's Paradox is deflated.

1.5 Factors Which Affect Rates of Cultural Evolution

We would anticipate that civilizations would mature as a natural course of events. However, that may not be true. We must refrain from automatically assigning the science of Earth as the set of natural laws that govern other celestial bodies. The natural laws, or at least the understanding of them, are probably different in each alien civilization. It appears to the best of our knowledge that the science of earth is the science of our solar system. That assumption may point out how immature our civilization is; the science of the other bodies may be radically different from our own. We observe other celestial bodies and try to explain them based on Earth science. Unfortunately, that is what we know; therefore, it is to be expected that we behave in this manor.

The behavior of our civilization is well described by Thomas Kuhn in *The Structure of Scientific Revolutions*. As Kuhn observed, we Earthlings tend to use "normal science". That is that we labor diligently to match facts and observations to the existing theories or paradigms. This normal state is interrupted periodically by the occurrence of an anomaly, which creates crisis. The resolution of crises foster scientific revolutions which allow progress. We as a civilization may be faced with a need to break from "normal science" or more specifically mainstream science in order to progress at a rate that meets our needs.

There are many factors that can influence the rate of maturation of a civilization. They can be grouped in two categories. Those are environmental and intellectual. We will attempt to postulate some of those factors.

Let's address environmental factors first. Those are factors that are related to the natural resources available to a civilization. An alien civilization could have the good fortune to have available to it far greater stores of natural resources than do we. An alien civilization might have resources much more suited to space travel than those available to us. Obviously, it is also true that some other civilizations could be less fortunate than Earth from an environment prospective.

Intellectual factors are potentially more diverse than are environmental factors. Another civilization could have come onto the scene at a radically more advanced level, perhaps at a level where we will be in another few hundred or even thousands years of development. They would already possess technologies that we are not yet dreaming about. The converse is true, as difficult as it seems to believe, that another civilization could have been even less developed at its birth than ours. It is important to remember that while Earth's population is about three million years old, we are really only a few hundred years old from a technology development perspective. Earth's technology is driven by the contributions of a few individuals such as the person responsible for the capturing of fire, the person who developed the wheel, the person who developed the written alphabet, Aristotle, Archimedes, Einstein, Oppenheimer, Hawking, Newton, Lavoisier, Fermi, and countless others. The parallel or superior intellects in other civilizations may have come along significantly earlier or later in the development of their culture. A third intellectual possibility is that the average intelligence of the masses is either considerably higher or lower than of Earthlings. This too would affect the rate of development. More important probably than the average intellect, is the existence of a few extremely intellectual beings with the opportunity to develop and be heard. Many of our technological advances have been the result of exceptional interpretation of observations and findings by a few exceptionally intelligent people. The efforts of one exceptional intelligence often advances our technology faster than the efforts of thousands of average contributors. Average contributors at best make modest improvements on the work of those exceptionally intelligent people as the masses typically practice "Normal Science."

1.6 Compatibility of Technology and Environments

An important question can be raised about the viability of an alien civilization's technology in the new host's environment. If "Normal science," as defined by Earthlings, pervades the universe, we would expect the technologies of all civilizations to be compatible with all other environments. The statistical analyses presented earlier lead us to the conclusion that this is unlikely to be the case. There are just too many potential civilizations to believe that the laws of science are equally understood by all or perhaps even most civilizations.

It should be anticipated that the intruder has verified by way of unalien tests or simulation that his technology is viable in the environment that it intends to invade. The question becomes most important in the case of insufficient time to perform such tests, the invader's lack of thought to perform such tests, or an error(s) in the simulation.

We would assume that technology is compatible with the new host's environment. However, there are cases where it is likely that the alien's technology will not work for a sustained period of time. Some examples of potential incompatibility follow.

Incompatibility Case 1: We were to send our metallic devices into an atmosphere with a high Fluorine or Chlorine content. It is possible that other lifeforms could survive in that condition. However, many of our materials would be quickly eroded to the point of losing structural integrity. Similarly, the aliens might require something analogous to an Earthling's Oxygen supply.

Incompatibility Case 2: An Alien fleet arrives on Earth in one or more ships built of a material also alien to Earth. There is the possibility that the oxygen, nitrogen or the combination of the two in our atmosphere would aggressively erode that material to the point of making it structurally inadequate. Or there simply might be none of it available in raw form in our Solar System for the aliens to implement repairs with.

Incompatibility Case 3: An alien has a light-based weapon that relies on speed of travel through its native environment to be effective. The rate of travel through Oxygen and Nitrogen could be lower and render it useless here. It is also possible that the weapon be designed to stun its victim; however, the speed of travel through our environment could increase its potency to the point of making it lethal.

Incompatibility Case 4: Bacteria unknown to the intruders in our atmosphere or that of a host alien civilization would digest the materials of an invading culture.

Incompatibility Case 5: The density of the host's atmosphere might be in a range that prohibits flight because of the density of the invader's spacecraft materials.

There are many possibilities for incompatibility of alien technologies within host environments. However, there is perhaps at least as high a probability that the technologies of aliens, while radically different, are infinitely compatible with the environments of their chosen hosts. These few scenarios were included to fuel the reader's imagination and to make the point that technological and biological compatibility will impact the probability of alien interaction. More advanced civilizations should realize if they are compatible with lesser ones, perhaps. Therefore, compatibility is a factor that might be added to the Drake Equation discussed in previous sections. We hope that you will spend the time to conduct the thought experiments to consider other possibilities.

1.7 "Putting All Our Eggs in One Basket!" a digression to a debate between the future Dr. Travis Taylor and Dr. Carl Sagan

The story that follows is an actual debate that was held between a young Travis Taylor and the esteemed Dr. Carl Sagan during a rainstorm in a parking lot in Huntsville, Alabama after a lecture by Dr. Sagan during the summer of 1986. This story was added to this textbook after some discussion between the authors. It was decided that the story would help support the main theme of this text, which is that we cannot count on the first alien contact to be a happy one. The authors all agreed that the story would come across best if told from the first person by Dr. Taylor himself. So, the rest of this story is as presented by Dr. Taylor.

I had always wanted to be a scientist, since the third grade. I recall watching all of the *COSMOS* series on public television with great awe and admiration for Carl Sagan. I was an avid fan and a member of The Planetary Society. In fact, it was one of the episodes on radio telescopes that inspired me to build my own radio telescope as a high school science fair project. I began the project in the summer before my senior year and completed radio observations of Cygnus A just in time for the local science fairs and science paper competitions. My project went all the way to the International Science Paper Competition where I took sixth place in my division. I felt that I owed it all to Dr. Sagan. I even wrote a letter to The Planetary Society stating such.

One of the prizes I won was a summer job at the Redstone Arsenal in Huntsville, Alabama. My other big interest at the time was in lasers. Fortunately, the job that I won was working in a Directed Energy Research Laboratory with the Army. I gained a lot of technical experience in that lab.

About midway through the summer Carl Sagan came to town and presented a lecture at the Wernher Von Braun Civic Center. I was thrilled and like a kid at Christmas due to the prospect of getting to meet Dr. Sagan. So, I grabbed my copy of *Contact* and forced several of my friends to go with me to the lecture.

With hopes that Dr. Sagan would speak about great things like the cosmos and the immenseness of the universe and the Search for Extra-Terrestrial Intelligence, I was soon given my first taste of politics and the vast gulf between the left side and the right side of politics in America. Dr. Sagan spoke very little of great things. In fact, his whole lecture was designed to belittle and ridicule any and all scientists working on the Strategic Defense Initiative (SDI). He made continuous wisecracks and very unfounded statements as though they were scientific fact, and they were not. His final conclusion was that there would be no hope for peace between the Soviet Union and the United States if SDI were continued. He continued to state as fact that the only hope would be for the U.S. and the Soviets to join on a manned mission to Mars. This joint manned mission was our only hope to end the Cold War.

I could not believe what I was hearing. Perhaps he had just missed the point of SDI. I had to ask him some questions. So, I jumped up and ran to the line to ask questions. The line was quite long. Unfortunately, the local community played softball with Dr. Sagan. I was surprised by this in that none of the local scientists I had met ever played softball with me. Perhaps they were afraid of the legend of Dr. Sagan. His reputation as a scientist was well boasted and all us hicks from Huntsville ever did was build the rockets that went to the moon. What did we know?

My chance to speak was almost there. I was next in line. The fellow in front of me stepped up to the microphone and asked, "Dr. Sagan, have you seen Johnny Carson lately and how is he?"

What a wasted opportunity I thought. After Sagan chuckled a response to the fellow, he replied that that was all the time he had for questions and that he would be signing autographs for a few minutes. He also responded that he had hoped to get at least "one good question" from the crowd and that he had not. My grandmother could have taught him a thing or two about catching flies with honey as opposed to vinegar!

Since I did not get to ask my question before, perhaps I could ask him while he autographed my book. At least that was my hope. It did not work out that way. I had to hand my book to a Civic Center guard who in turn handed my book to Dr. Sagan up on the stage. He scribbled in it and it was passed back to me. I did not get to within twenty feet of him. So I tucked my book under my arm and moped to the door where my friends awaited me.

It was pouring out and we were going to make a run for the car. We dashed to the car and everyone piled in. One of my friends pointed out that there was a limousine about twenty feet away from where we were parked. And believe it or not, here came Dr. Sagan walking under the umbrella being held out for him by his chauffer. I had one more chance.

I stepped out of the car and back into the rain. As I cautiously approached Dr. Sagan and his chauffer I announced my presence. I did not want to startle them off. Celebrities probably get that sort of thing far too often.

"Dr. Sagan," I said. "I think I have the one good question for you!"

The chauffer opened the rear door of the limousine and Sagan turned to look at me.

"Okay, you have ten seconds," he replied.

"Well, uh, you claim that a joint manned mission to Mars is the only way that we can end the Cold War and that SDI is a big waste of time, effort, and money. Especially since you believe that it is technically impossible. Isn't that putting all your eggs in one basket? What is wrong with doing both? I really believe that it is too early to say that SDI is technically impossible. And what if it has the side effect of bankrupting the Russians? SDI might work in a different way." I was very wet, but I was proud of myself. Then he responded.

"Nonsense. SDI is an absolute waste of money and until we grow up and venture into the stars as a species together, we will never know peace. I have to go now your time is up." He paused and then finished with, "I am still waiting for that one good question." He closed the door and his limo drove away.

I stood there in the rain destroyed. A childhood idol and hero turned out to be a total jerk. I actually used much harsher language. I stood there for a while until my friends yelled for me to get in the car.

The next day I wrote a long letter to The Planetary Society explaining and demanding that such one sided thinking was not healthy for the American way of life. And on top of that it was obvious that there was a definite political bias against Ronald Reagan and his plans. I stated that I joined The Planetary Society to learn about space and planetary science not to become a pawn of a political advocacy group. I sent the letter.

I got a typed response from the director of The Planetary Society, Louis Freedman, stating that

"...while we have great respect for Dr. Sagan and his views, his personal beliefs and statements do not necessarily reflect the views of The Planetary Society. Thank you for writing and please consider donating to the..."

I promptly quit the organization and showed both my letter and the response letter to my then boss and mentor at the Army. He laughed at first until he saw how troubled I was. Then he became a true mentor to me and I will never forget the conversation we had.

"Well, I wish you would have shown me this letter before you sent it," my mentor Dr. Honeycutt started. "I could have saved you some trouble. Son, the country is split between what I would call Utopians and Realists. The Utopians believe that being able to defend one's self is not necessary because we are all humans and should be able to live happily ever after together. Us Realists on the other hand have had our lunch money stolen from us by the school bully. And the Realists know that punching the bully in the nose will go a long way in preventing having your lunch money stolen again. What you have to learn is that the country is about half Utopians and about half Realists. You also need to decide which one you are. If you are a Utopian you will agree with Dr. Sagan and continue chasing a one-world government through this manned Mars mission. If you are a Realist you will take your talents and help us punch the bully in the nose."

I stayed with the Army and worked for the next twelve years on high-energy laser weapons concepts. I watched as the tests of the SDI demonstrators failed and had some success. I was a Realist by mentor's definition as SDI became the Ballistic Missile Defense Organization and then the Space and Missile Defense Command, and then the National Missile Defense and so on.

I also was a Realist by my mentor's definition when the spending on SDI was more than the Soviets could match and/or counter. I continued to be one when the Soviet Union went bankrupt due to out of hand defense-spending requirements. SDI had succeeded!

I was most definitely a Realist as defined by my mentor, Dr. Thomas E. Honeycutt, when he died of cancer. He had spent his life as a "Realist" and trying to use his brilliant abilities to help maintain the American way of life. I am still a "Realist", and in Dr. Honeycutt's honor, I will always be!

The point of this anecdote is to express that there are differences in opinions, which elicit healthy discussion and debate. In fact, the authors would be all for a joint manned mission to a space destination while at the same time agreeing with the value of SDI. The SETI community would have most people accept the concept that all advanced alien cultures have evolved beyond the need for war or even malicious or self-serving intent. Dr. Sagan was a big proponent of this belief. Why else would the SETI community be looking mostly at one set of frequencies and expecting an alien civilization to have figured out that that is where we are looking for a signal? And, that the aliens were broadcasting a nice message including the *Encyclopedia Galactica*. *Contact*, the book and movie, is a good example of this philosophy. This philosophy is a rather optimistic concept.

Science forces one to admit that the *Contact* scenario is a possibility – in fact a preferred possibility. On the other hand, science would also suggest from the statistical arguments given in earlier sections of this chapter that a much more malevolent and malicious scenario might exist. However large or small the probability is, the probability of an ET attack *is* finite and does exist!

It makes good sense to prepare for the worst and hope for the best. SDI is a good example. We prepared to defend against a missile attack and hoped that we would bankrupt the Soviets and therefore prevent a war. We are still preparing and still hoping but now against other possible earthbound enemies. We should also prepare, yet still hope, for the non earthbound possibilities. There are Near Earth Objects (NEOs) and comets that might cause us problems one day. There are major disasters caused by volcanoes, category 5 hurricanes, massive earthquakes, and tsunamis, and countless large scale natural phenomena. We are ill prepared for those and we know they happen. We should think now about our civil and global defense preparations for potential disasters and onslaughts of all possible types and scenarios. Hopefully, our preparations will be as brilliantly successful at meeting our objectives as Ronald Reagan's SDI concept has been.

1.8 The Central Limit Theorem

The central limit theorem is a theorem about statistical trends over large sample space. The theorem simply states that the sum of a large number of independent observations from the same distribution has, under certain general conditions, an approximate normal or Gaussian distribution – this is the so-called "bell curve" and it follows a mathematical function known as the Gaussian function. The approximation steadily approaches a Gaussian function as the number of observations increases and if infinite observations are made the curve *will* become exactly Gaussian. The central limit theorem is considered the heart of probability theory, although a better name might well be the normal convergence theorem.

It has been empirically observed that various natural phenomena, such as the foot size of individuals, types of stars in a galaxy, color of jelly beans in a jar, atmospheric molecular motion, rolls of a dice, economic trends, biological divergence, and most all random phenomena, follow approximately the so-called normal distribution – a Gaussian function. A plausible explanation is that these phenomena are sums of a large number of independent random effects and hence are approximately normally distributed by the central limit theorem.

The central limit theorem makes three statements:

1) the mean of the sample is equal to the mean of the population from which the sample was drawn
2) the variance of the sample is equal to the variance of the population from which the sample was drawn divided by sample size
3) If the population is distributed normally as described by a bell shaped curve, the sampling distribution of means will also be normal. If the sample is not distributed normally, the sampling distribution of means will approximate normal as the size of the sample increases.

The obvious question now is, "What does this have to do with ET invasions or the likelihood of ETs?" We propose here that the central limit theorem is the basis of most unknown and random variables in the universe and therefore is the best means from which to extrapolate statistical possibilities. There are many who would argue the plausibility of other intelligent races from religious, statistical, or other venues without taking into account the central limit theorem.

Most things in the universe follow the normal or Gaussian distribution and are known as random Gaussian variables (RGVs). Stellar types in galaxies follow this Gaussian type distribution and therefore have lead astrophysicists to suggest that various star clusters have similar origins since they follow the same type of mathematics.

It takes very little leap of faith if any, merely mathematical deduction, to realize that if star systems follow such a mathematical basis and planets are byproducts of those stars, then planetary formation must also follow this central limit theorem. Actually, they should most likely follow the central limit theorem simply due to the large number of planets in the sample space. Further deduction from the planet types to the types of lifeforms that might develop on these planets should still follow the same type of function since the variables are all closely coupled. The central limit theorem rules the day in our universe.

From the logic of the central limit theorem we can extrapolate that the types of creatures and their motivations that could exist in our universe should follow a Gaussian distribution. The question is what is the mean or average type of lifeform. Figure 1.14 shows a typical Gaussian distribution that is applied to the description of lifeform types in our universe. The average or mean type of lifeform is most probable to occur and those that are most benevolent and hostile towards those lifeforms are located in the less probable regions on either side of the mean. The problem with this model when applying it with the knowledge that we have is we do not know what the mean species is like. Are we the average type of species? Who knows? Assuming that we are – and there is no basis for this assumption – then there are more species similar to us than dissimilar. However, if we are beyond a standard deviation and are in the "wings" of the curve then there are many more species very dissimilar from us.

Even if we are the mean type of species, due to the vast size of the universe and the large numbers involved with the number of species that should be available, there is still a significant number of species that might be very detrimental to our existence.

There are those who argue that we are alone or rare because we will eventually blow ourselves up or destroy ourselves and that is why we aren't hearing from other species. We haven't done this yet and again this is a variable in a large sample space which suggests that the central limit theorem applies again. So, if the mean is that most people blow themselves up there would still be species beyond one standard deviation on the curve that do not. Or if the mean species does not blow themselves up, that would suggest that the self annihilators exist in the "wings" of the curve beyond a standard deviation. Either way, there are still a significant number of species that would not self annihilate via the central limit theorem.

Another factor to consider are the various physiologies of alien lifeforms as well as the geological make-up required for them to form. While there is considerable disagreement on the method of intelligent life arrival on Earth, there is little room for argument that it is sustained because of favorable geologic conditions for our type of life to develop. The favorable geologic conditions, according to Simeon Winchester in his treatise on *Krakatoa*, are Earth's gravity and its distance from the Sun. Of course, exobiologists have developed many theories and suggest many other driving factors. Books like *Rare Earth*, for example, would lead us to believe that ET life is near impossible to occur. Again, there is no way to determine where on the Gaussian curve we fall, so such conclusions are purely speculation.

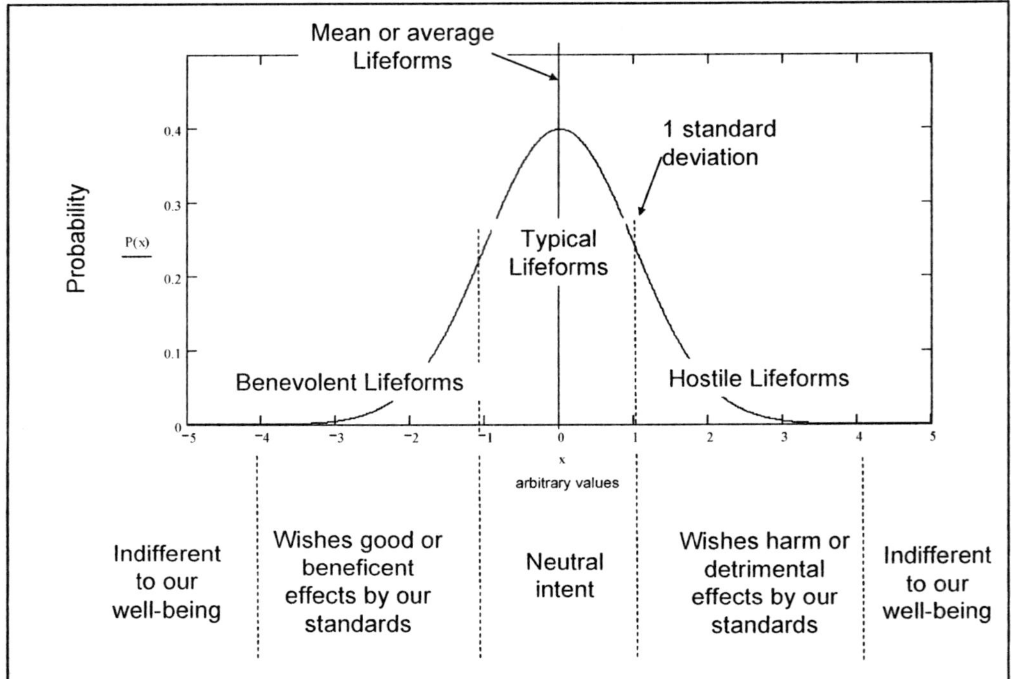

Figure 1.14 Central Limit Theorem suggests all types of aliens. Note that this graphic assumes (arrogantly) that humans are the mean or typical lifeform.

On the other hand the Earth's gravity is such that it prevents precious things like water, oxygen and carbon dioxide from escaping our atmosphere. The trapped atmosphere blocks out harmful things like UV, really high energy gamma, and cosmic rays. Human bodies are comprised mostly of water and water is crucial to our existence. Oxygen also is required for the continuance of life *as we know it*. Animals take oxygen from the air and use it to carry on processes and release carbon dioxide into the air as a waste product. Plants take carbon dioxide from the air and use it to produce food for themselves while releasing oxygen into the air as a waste product.

The Earth's distance from the Sun also has a significant influence on the water found on Earth. The Earth is advantageously close to the Sun so that the water is not found only as ice. It is far enough away from the Sun that the water is not all found as steam and therefore, inaccessible to sustain life.

Are all of these factors average to planets across the universe or are they rare? Where do they fit on the Gaussian curve? One thing for certain is that they do fit on that curve somewhere due to the central limit theorem again. Therefore, there must be multiple such planets; *how* abundant is the question.

ET must also exist because of favorable geology on his home celestial body. ET may in some cases require similar geologic factors if he too is dependent on the same compounds and conditions as we Earthlings. Logic dictates that some number of ETs probably require geology and chemicals different from those we use to sustain ourselves. It also follows that many ET's are probably different from one another. This allows a much larger number of celestial bodies to serve host to forms of intelligent life. And this requires us again to consider where and what are the mean or average types of alien lifeforms in the universe as the central limit would dictate.

1.9 Chapter Summary

This chapter was designed to offer the reader a different look at the possibilities that might one day confront mankind. In section 1.1 a discussion of the Drake Equation is given that shows that there is actually a range of solutions to the Drake Equation that has startling endpoints. As is often taught in the general astronomy community, one end of the range is that the odds for other alien life are very slim. However, the other end of the range is that Earth is visited every few thousand years by aliens. Even a midpoint between this broad range is still a finite probability of alien visitation – the key emphasis here is on the phrase finite or nonzero probability. All solutions except the zero solution suggest that the probability of alien life elsewhere is indeed finite – how big or how small is the only question remaining.

Section 1.2 introduces the point that we are currently broadcasting our whereabouts to the rest of the galaxy. That might be a very bad idea. A more appropriate defensive posture might be for us to "hide and watch" the rest of the universe until such time that we are technologically capable to defend ourselves from possible alien invasions.

The so-called Fermi's Paradox is discussed at great length in Section 1.3. Multiple population models are discussed as well as details of why Fermi's Paradox is a scientific blunder. It is hoped in this section that the paradigm of Fermi's Paradox as an accepted explanation for humans being alone in the galaxy is completely debunked.

Section 1.4 discusses the distance barriers between star systems and how to modern day scientists it is often considered an impossibility to travel the vast interstellar distances. In this section we introduce the Kardashev scale for characterizing alien species based on their capabilities of power manipulation. We also offer a more readily applicable method for understanding advanced alien civilizations based on how quickly they can travel interstellar distances.

Section 1.5 discusses a modeling approach for understanding rates of cultural evolution. A linear systems model of a cultural system (population model) is given that emphasizes the complexity of such a system. Truly understanding such a system would require an intensive research effort.

Even if there are aliens out there that intend to attack other civilizations, there are possible factors such as mere species compatibility with the atackee's world. Section 1.6

briefly offers some examples of how some technologies would interact poorly with certain environments. Realizing such incompatibility possibilities may exist, tells us that another probability factor for this phenomenon should be considered in any statistical analysis of possible alien invasion.

A brief digression into a debate between the future Dr. Travis Taylor and Dr. Carl Sagan is given in Section 1.7. There are multiple points to this section – some subtle and some not so subtle. The major emphasis is that as a species we are accepting from the "popular science" community that the probability of an alien invasion does not really exist. Statistics argue against that position. The probability of an invasion is as equally likely as not being invaded when the limited knowledge at mankind's disposal on the subject is considered. Perhaps as we understand the makeup of the universe better in the future we will understand the probabilities better. But for now, ignoring the probability is in essence "putting all our eggs in one basket" that the aliens either will not show up or if they do they will have "evolved above war" as is so often suggested by the Sagan philosophy.

Finally, in Section 1.8 we explain the Central Limit Theorem and why it applies to the topic of alien invasion. The probability theorem tells us that most things in this universe are not an either/or situation. There are degrees of either and degrees of or called standard deviations. For humanity to be the only life in the galaxy, let alone the universe, would make this an extremely unlikely universe following the theorem.

2

Warfare with ET – Human versus Alien Weapons, Tactics, & Strategies for Humanity's Defense

In the event of an alien invasion on a global scale mankind would have to implement every weapon, tactic, and strategy at its disposal. Students of military history and warfare might initially be inclined to think that warfare is warfare universally and historical theory can be applied to any scenario. This is true in general terms and is a good starting point for developing a plan of action. However, it is likely that completely new models of warfare will be required that account for alien concepts that we have yet discovered. Just as warfare modeling had to be adapted when aircraft technology reached capabilities that made them useful weapons so will our present models need to be adapted to alien capabilities perhaps like teleportation devices.

In this chapter we will discuss some of the basic principles of war and how they fit or need to be adjusted to allow for alien invasions. And we will discuss the tools of warfare that might be implemented in the event of that invasion. We will not provide an exhaustive list of technology, as that would be near impossible at this point, but instead we will give a few examples of specific ones that might be effective. We will also cover possible modes of a global invasion that need to be addressed such as civil defense, asymmetric warfare, and various potential "force multipliers".

2.1 The Basic Principles of War Adjusted for ET Invasion

In the book *How To Make War A Comprehensive Guide to Modern Warfare for the Post-Cold War Era Third Edition* James F. Dunnigan gives a very good description of the twelve major components of war for the "armchair general". We list here these twelve components but with definitions adjusted to meet the needs given an alien threat rather than a post-Cold War era threat. The basic principles of war are:

1. **Mass.** "Get there first with the most." This point holds with any scenario. Victory usually is gained by the side that can amass the largest fighting force on

the field of battle. A common quote from combat commanders is, "The more you use, the fewer you lose." However, extremely technologically superior troops might alter this philosophy. How would individual alien troop armor or shields that could withstand a direct hit from a rocket propelled grenade change this principle?

2. **Unity of Command**. Multiple branches of the combat force (ground, air, sea, space) must operate toward the same goal, execute the same plan, and execute prearranged actions while maintaining a cohesive chain of command. This principle has been followed well by American forces historically and even by allied coalition forces. But in the event of an alien invasion, global coalitions might be required between allies and foes with opposing agendas. How can we follow this principle in the global invasion scenario?
3. **Maintenance of the Objective**. Dunnigan defines this as "choosing a reason for being on the battlefield and sticking with it". This means not changing the reason for particular battles midcourse. A good example of this is the ground campaign in the second Gulf War. The objective was to move the U.S. and coalition ground forces across the country as rapidly as possible while taking out all Iraqi warfighting capabilities along the way much in the same way the Israelis did in the Arab-Israeli wars in the 1970s. By not stopping (in the first few days) along the way to perform "mopping up" tasks the forces were able to cover the country rapidly and take out many more of the concerted Iraqi forces in a short period of time before they could react. How does this apply in an alien invasion scenario? One possibility is to consider defining the threat as subsets and attacking one subset at a time (i.e. attack troops in one strategic geolocation first, then move on). How else? This needs further investigation as the global aspect alters the paradigm.
4. **Economy of Force**. All fronts cannot be fought at once. Hitler learned this the hard way. Dunnigan calls this "putting all your eggs in one basket", which is aptly appropriate. As the objectives are defined and broken into subsets as described above, multiple force requirements will be needed. Will there be enough for multiple campaigns globally or will our forces be spread too thin? Hitler also implemented large reserve forces which prolonged World War II. Perhaps this is a tactic that will come in handy during an ET invasion.
5. **Flexibility**. Maintaining flexibility in planning, tactics, strategies, and action is according to Dunnigan "common sense". Since we will not know much about the alien threat and how they make war, we will have to be very flexible in adjusting to their attacks – their alien attacks. We must keep in mind this flexibility but not at the cost of "Maintenance of the Objective". The two must be traded appropriately. If initially the objective is to defend a region, but it is discovered later that an all out attack on an alien headquarters/hive/node/whatever would have a high probability of winning the war, do we change the objective midcourse?
6. **Initiative**. Dunnigan describes this by suggesting that surprise is simply the disparity in initiative between two forces. An example of this is the U.S. airborne

troops on D-Day versus the German commanders. The German combat commanders were afraid to take action without orders from high command while the American troops were given the free rein of taking the initiative and doing what seemed necessary on the battlefield scale to win the war if cut off from higher command. Not fearing reprisal from higher command if they took action, enabled the U.S. troops to, in small groups, pick targets of opportunity and take them out without specific HQ approval. In a global invasion, it is likely that initiative will be what separates the surviving forces from the extinct. Communications globally would most likely be lost and forces would need to react locally and immediately without the time delay of orders from headquarters.

7. **Maneuver**. You must move your troops around. Depending on the nature of the global alien threat, actually moving them to *where* is an issue. In L. Ron Hubbard's *Battlefield Earth* series, Earth's forces are wiped out by aliens in seventeen seconds – little room for maneuvering there. But in *Independence Day* the air forces of the world kept running and hiding from the alien saucer ships until they figured out a plan and then they acted in a concerted effort with a solid objective. What other aspects of maneuvering are pertinent?
8. **Security**. To paraphrase Robert A. Heinlein, "a secret plan must be that, a secret". If the aliens become aware of our one good plan to defeat them, how effective would it be?
9. **Surprise**. Surprise is always a force multiplier and is typically safer and more successful. From an alien invasion standpoint, they will probably surprise attack us. How do we prepare for that?
10. **Simplicity**. Warfare is a wild chaotic violent undertaking that makes adhering strictly to planning extremely difficult. So, the more simple the plans, tactics, and strategies are the easier they will be to follow. Fighting on a global invasion scale compounds this principle and makes it even more difficult yet more important to follow.
11. **Morale**. Once the aliens show up and destroy the majority of Earth's defense capabilities the remaining forces will quickly lose morale. We will need to find motivators to keep the remaining forces mentally and physically able to carry on. This may be as simple as a "Get off our Planet!!" battle cry or it might prove to be our greatest hurdle. A demoralized army cannot fight well.
12. **Entropy**. Perhaps after the initial alien invasion the war will settle down to a steady grind. What troops are left are caught up in a day-to-day fight against the occupying ETs. At this point the war is easier to model with predictability and implementing strategies will become more important than simply reacting with survival tactics as during the initial invasion. The capability for humanity to understand and manipulate the entropy after the initial alien onslaught might be the key to winning the war – of course, that assumes the aliens do not wipe us out in the first seventeen seconds.

Understanding What the War Might Look Like

Assuming that the twelve basic principles of warfare are still, for the most part, valid under an alien invasion scenario, we can model and simulate such combat scenarios in order to offer the military strategist insight into needs, tactics, strategies, and planning required for preparing for such an invasion. Or, at least we can model what the "steady grind" of the war might look like after the initial invasion with some confidence - although, simulation of localized ambush is well understood with modern combat modeling techniques, so we might be able to apply that knowledge to global invasion with some effort.

We will discuss here, briefly, a model for force-on-force warfare simulation that can be used to do initial analysis of the ET invasion scenario. The model discussed herein is based on the rate-based deterministic differential equation models developed originally to simulate and model biological species in competition and niche crowded environments similar to those discussed in Section 1. This approach was first used by F. W. Lanchester in 1914 to simulate and study warfare. It is the original Lanchester efforts that are the basis of all of the aggregated combat force attrition simulation models accepted today by the Department of Defense. Our present simulation tool is a modified and updated deterministic system that would be classified as an "enriched" Lanchester-type model.

Figure 2.1 gives a graphical representation of the model's capability to simulate combat scenarios. The graphic shows the various components of warfare as implemented in the enriched Lanchester model. We will assign the Blue force as human and the Red force as the alien invaders. Of course, no model is completely accurate, but the insight gained from running such simulations is beneficial in understanding the general flow of a combat scenario.

Using the simulation model shown in Figure 2.1, each of the force arrows is represented mathematically as a sum or a product (depending on the situation) of coefficients that represent particular force capabilities and multipliers such as new armor, weapons, and other effectiveness capabilities. A new concept or technology can be represented by inserting a new coefficient into the simulation. The weight of the coefficient is based on the literature/archival data and/or a "back-of-the-envelope calculation". The robustness of this model enables us to simulate weapons that are science fiction in description, such as shields, by inserting a coefficient multiplier into the simulation. In the case of things like shields and other science fiction concepts, the magnitude of the multiplier is arbitrary and little more than a guess. Although, the model can be run quickly and many times for different values of the shields multiplier and therefore show us a range of outcomes due to various levels of shield strengths.

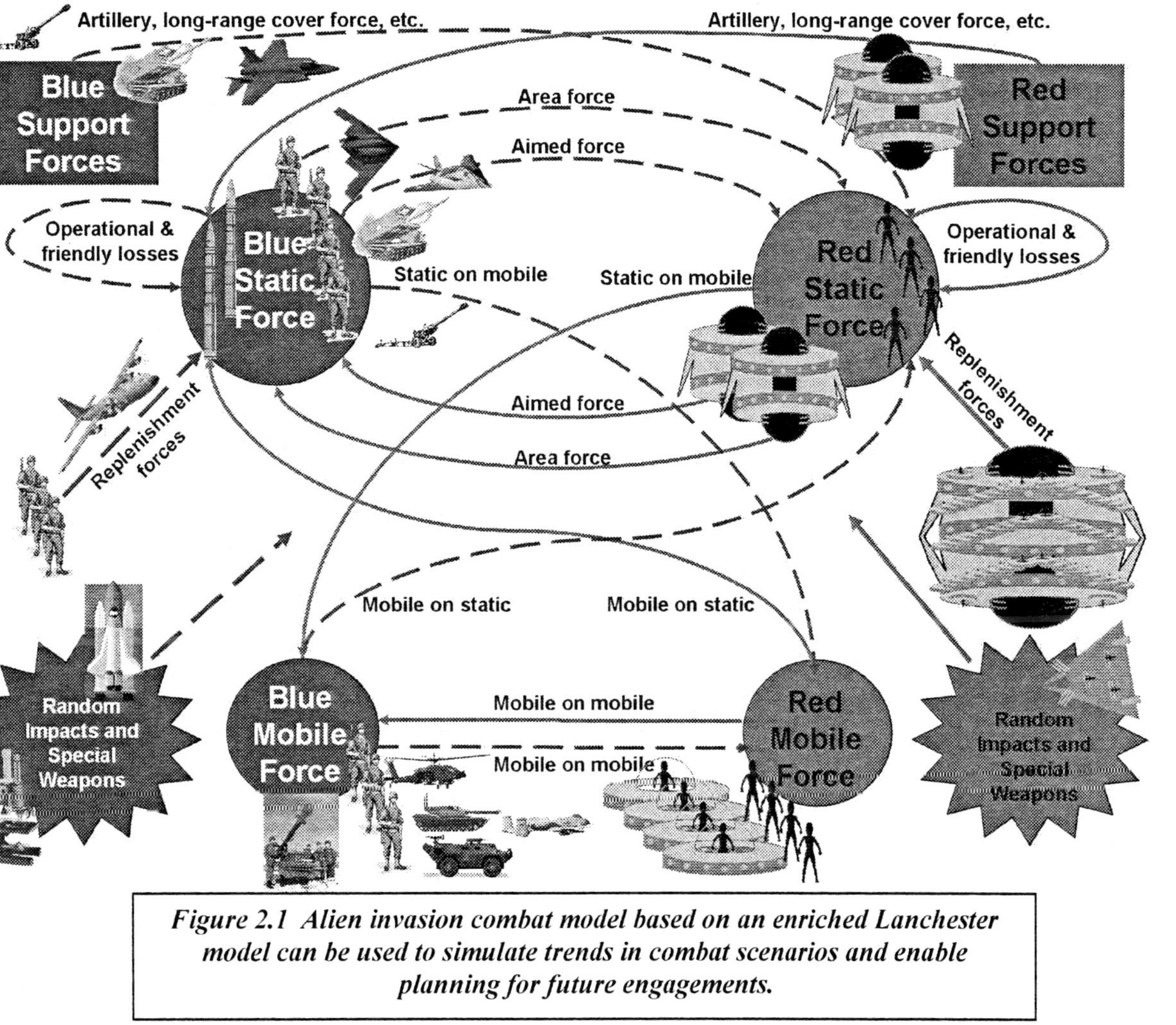

Figure 2.1 Alien invasion combat model based on an enriched Lanchester model can be used to simulate trends in combat scenarios and enable planning for future engagements.

The force-on-force model depicted in Figure 2.1 is mathematically represented as a set of two coupled differential equations that are the rates of change of the Red and Blue forces due to interaction with each other and themselves as a function of time. The two equations are

$$\frac{d}{dt}\mathrm{Red}(t) = -a\ \nu\ \mathrm{Blue}(t) - A\ \mu\ \mathrm{Red}(t)\ \nu\ \mathrm{Blue}(t) - \beta\ \mathrm{Red}(t) - \mathrm{SB}\ \mathrm{Red}(t)\ldots \\ -\ \mathrm{Ams}\ \nu\ \mathrm{Blue}(t) + \nu\ \mathrm{Bmm}\ \mu\ \mathrm{Redm} - \mu\ \mathrm{Asm}\ \nu\ \mathrm{Blue}(t) - \mathrm{DR}\ \mathrm{Red}(t) + \mathrm{rR}(t) \tag{2.1}$$

$$\frac{d}{dt}\mathrm{Blue}(t) = -b\ \mu\ \mathrm{Red}(t) - B\ \nu\ \mathrm{Blue}(t)\ \mu\ \mathrm{Red}(t) - \alpha\ \mathrm{Blue}(t) - \mathrm{SR}\ \mathrm{Blue}(t)\ldots \\ -\ \mathrm{Bms}\ \mu\ \mathrm{Red}(t) + \mu\ \mathrm{Amm}\ \nu\ \mathrm{Bluem} - \nu\ \mathrm{Bsm}\ \mu\ \mathrm{Red}(t) - \mathrm{DB}\ \mathrm{Blue}(t) + \mathrm{rB}(t)\ . \tag{2.2}$$

Where,

a = Blue force aimed fire coefficient
ν = Blue force multiplier
α = Red force operational/friendly losses
A = Red force area fire coefficient
SB = Blue force long range support
Ams = Blue mobile on Red static forces
Amm = Blue mobile on Red mobile
Asm = Blue static on Red mobile
DB = Blue desertions
rB(t) = Blue force replenishment rate
Bluem = initial # of mobile Blue forces

b = Red force aimed fire coefficient
μ = Red force multiplier
β = Blue force operational/friendly losses
B = Blue force area fire coefficient
SR = Red force long range support
Bms = Red mobile on Blue static forces
Bmm = Red mobile on Blue mobile
Bsm = Red static on Blue mobile
DR = Red desertions
rR(t) = Red force replenishment rate
Redm = initial # of mobile Red forces.

The two equations are solved numerically using a computer mathematics modeling software package such as Mathcad.

Figure 2.2 shows the outcome as a function of time for two exactly equal forces. Notice that when the forces are equal they tend to kill each other at equal rates so lines displaying the population of the Red and Blue forces overlay one another. By changing just the initial number of Blue forces to be ten percent less than the Red forces and leaving all other capabilities the same we see in Figure 2.3 that the larger force stays at a slightly higher value but still decays along a similar trend as the slightly smaller force. Figure 2.4 shows when the Blue force starts with fifty percent fewer troops. Note the difference.

Figure 2.5 shows what happens when we give the Red (alien) forces a force multiplier five times greater than the Blue (human) forces. All other coefficients are set equal. The greater force multiplier could represent any technological advancement such as precisely pointed high power directed energy weapons, ray guns, shields, or any other concept.

Figure 2.6 shows the Red force with a force multiplier of five times the Blue force but the Blue force initially starts with twice as many troops.

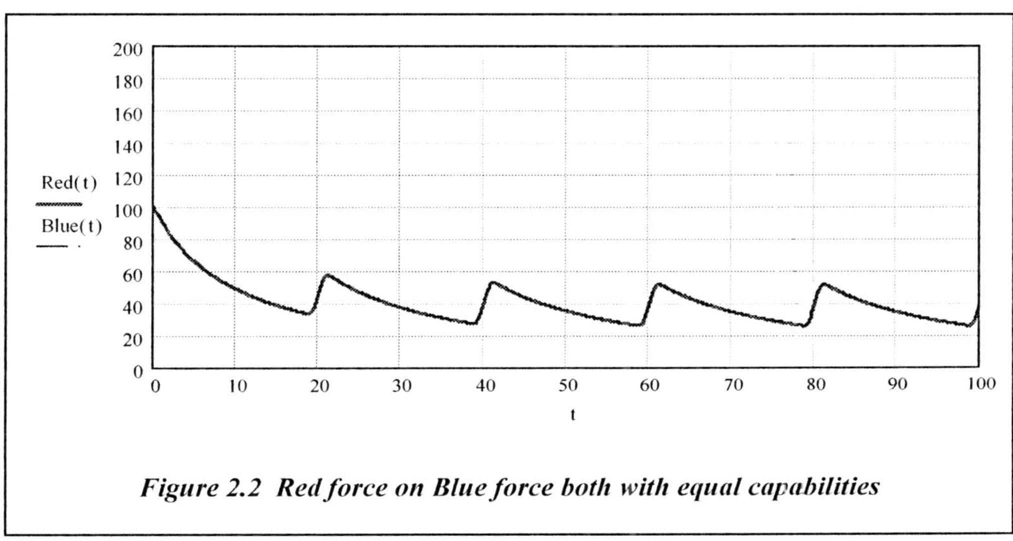

Figure 2.2 Red force on Blue force both with equal capabilities

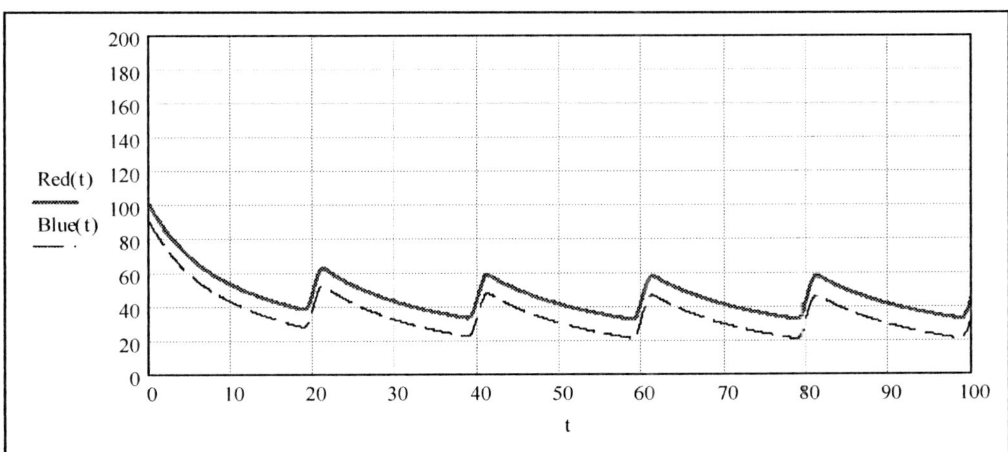

Figure 2.3 Red force on Blue force both with equal capabilities when Blue starts with 10% fewer troops

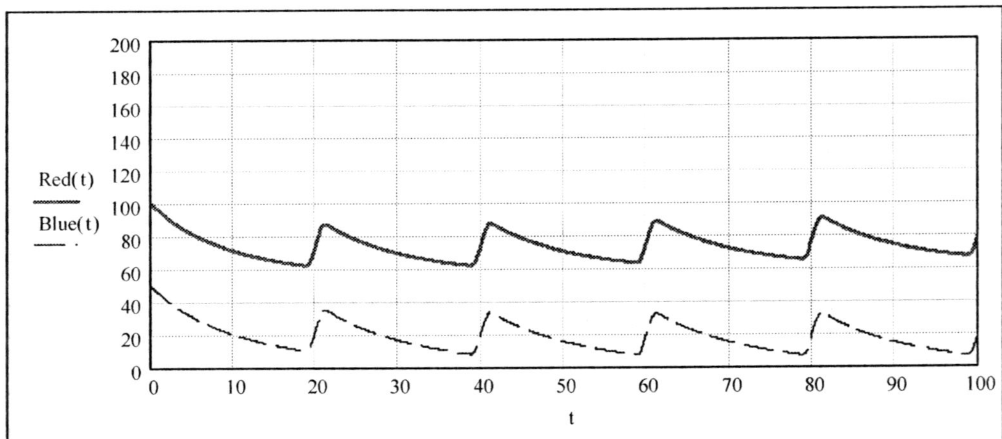

Figure 2.4 Red force on Blue force both with equal capabilities when Blue starts with 50% fewer troops

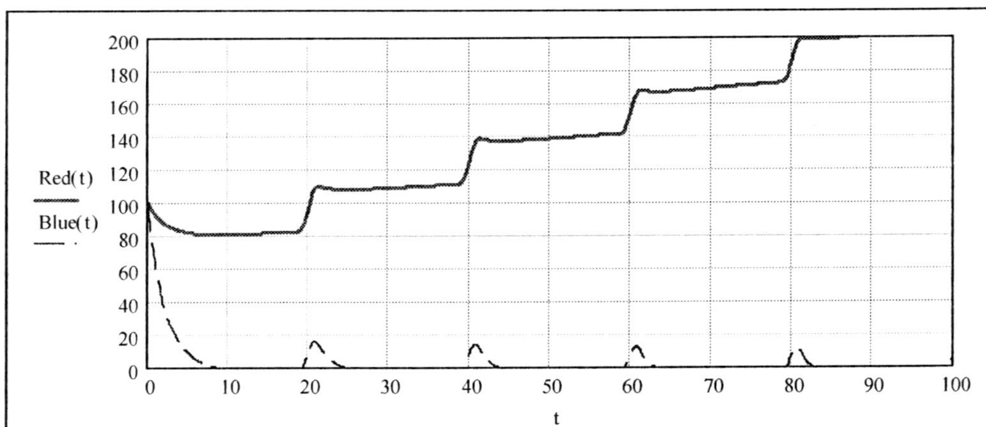

Figure 2.5 Red force on Blue force both with equal # of troops when Red force has a multiplier of 5 representing advanced technologies

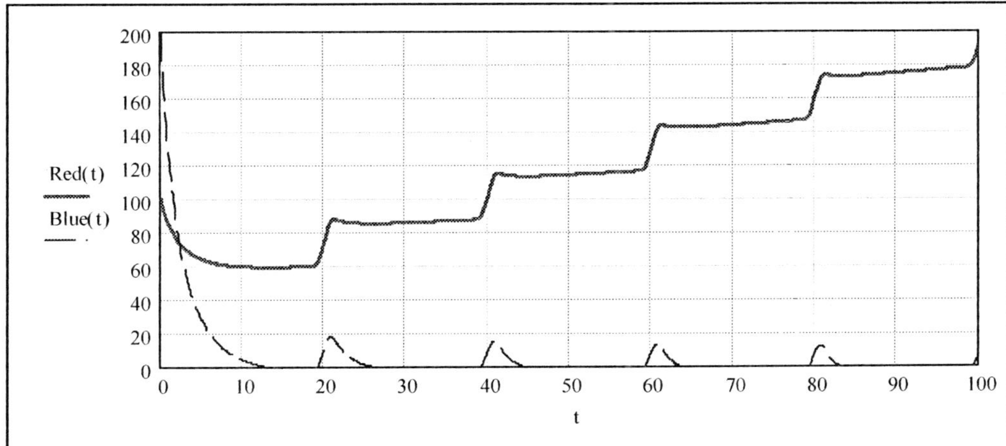

Figure 2.6 Red force on Blue force. Blue has twice the # of initial troops and Red force has a multiplier of 5 representing advanced technologies

In all of the simulations given here (other than the equal forces one) Red force wins. However, we assumed that both sides were able to replenish their troops with equal numbers of new troops at equal rates. Perhaps the alien invaders would not be able to replenish their troops as quickly or at all since they had to travel a long interstellar distance to get here – depends on what α level they are at and how fast they can travel. Figure 2.7 shows a simulation where the aliens never replenish their troops and the humans continue to replenish at the rates in the previous simulations. Also, the Red alien force has half the troops initially but a force multiplier of five.

In the case shown in Figure 2.7 where the Red force does not replenish their troops the results are quite interesting. Although, initially the Red force decimates the Blue forces, eventually, the new troops chip away at the alien forces and deplete their numbers. A slightly higher force multiplier for the Blue force or simply enough troop replenishments over time might enable humanity to defeat the alien threat.

The simulations given in Figures 2.6 and 2.7 represent an asymmetric war whereas the asymmetry is in technology differences. The simulation in Figure 2.7 is representative of humanity having an effective enough civil defense plans so that the civilian population can replenish the troops either as professional soldiers or as resistance fighters. If the Blue forces could hold out long enough with the periodic troop insertions then it is possible that the asymmetric war could be won. A good historical example of this type of asymmetric war working in favor of the lesser advanced culture is the Soviet-Afghanistan war in the 1980s. The Mujahideen were able to simply outlast the Soviets and eventually the Soviets gave up and pulled out of Afghanistan.

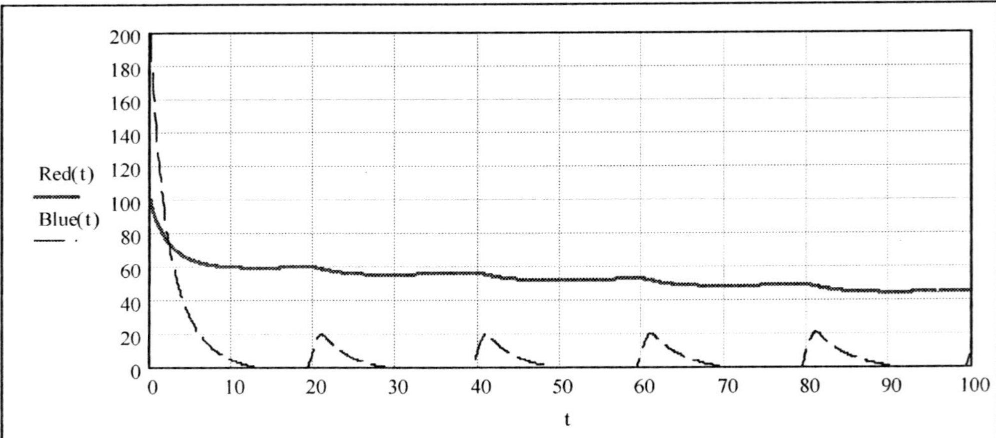

Figure 2.7 Red force on Blue force. Blue has twice the # of initial troops and continued troop replenishment. Red force has a multiplier of 5 representing advanced technologies and no troop replenishment. Note that eventually the Blue force numbers start increasing while the Red force numbers continue to decrease.

2.2 Civil Defense

What are the possibilities of large scale survival of humanity after an alien invasion? Could the civilian population survive? Well, of course the answer is a function of what type of invasion occurs and how well prepared we are for it. If we assume a global invasion, then we might be able to bound some of the possibilities.

The invasion is global, but like any attack the intensity will vary due to importance of targets. The aliens should have a minimum of radio signal and optical image intelligence of Earth. They will have detected the orbiting satellites and they should have a map of the globe of radio hot spots, population centers, and visible light hot spots. Figure 2.8 shows an image of the Earth from space at nighttime. The image is a mosaic of images collected between October 1, 1994 and March 31, 1995 by the National Oceanic & Atmospheric Administration (NOAA) and Defense Meteorological Satellite Program (DMSP). It is obvious from the concentration of light where the major population centers are located – the brighter the spots the larger the population center. If they have daytime imagery as well, then the aliens would have maps of our roads and infrastructure.

If the aliens took the time to understand the images and maps then they should be able to pinpoint the military bases, major airport facilities, vital roadways, and bridges. Unfortunately, our major military concentrations are most often surrounded by civilian infrastructure to support them. So, as these locations are taken out by alien weapons of mass destruction or invasion forces, it is highly likely that large collateral damage to the surrounding civilian population would occur.

***Figure 2.8** Nighttime satellite image of the Earth showing the lights of the major population centers. The image is a mosaic constructed via the NOAA/DMSP images collected from 1994-1995.*

The next logical target set is the population centers themselves. People are required to continue any defense action and even though the civilian populations are not initially military they would eventually become resistance fighters or lost to the alien attacks. Therefore, it is likely that the major population density centers are critical targets in an alien invasion.

But how would the general civilian population fare? The human race has had a number of opportunities to learn how to prepare for and recover from disasters. We have been faced with both natural and man-made events in the period from the beginning of 2000 to the end of 2005. Each has offered us the chance to learn something new and helpful. Unfortunately, as we review each we find that the issues, specifics, and requirements of each are different from others. The uniqueness of each is driven by geographic and local ideological differences.

The tragic and shocking attacks of Sept. 11 have raised fundamental questions about the shape and composition of future U.S. forces. An independent defense review published by the Center for Defense Information, and released just before the attacks, provides a roadmap for substantially restructuring the U.S. military to counter new threats in the first quarter of the 21st century. The review concludes that the U.S. military can, and must, be restructured to successfully undertake fourth-generation warfare against asymmetric threats — the kind of threats posed by terrorist networks. The study places a premium on the need for addressing personnel issues and doctrine, rather than hardware. It argues for boosting the cohesion and initiative-taking of U.S. troops, and asserts that the agility of America's forces should be enhanced by creating lighter, smaller and more mobile units. In contrast to the recently-issued Quadrennial Defense Review (QDR), the report suggests reductions in legacy forces to free up resources for transformation of other forces and for the other, ever more important, components of national security.

It is interesting that the responses made to 9/11 were military preparedness. Civil defense preparedness for such attacks still seems no better planned four years later than the day they occurred. Some will argue that the Department of Homeland Defense has put into place new efforts and better civil defense plans. Our thoughts on that are as follows:

- Do you know any more today about what to do than you did five years ago?
- Are there local terrorist attack shelters? If so, do you know where they are?
- Are there local anti-aircraft weapons to prevent the same types of attacks from happening? Where are they?
- Do the security protocols that were put in place at the airports post 9/11 really stop terrorists from getting on planes? Think about how you could take over a plane next time you get on one and you will most likely answer...NO.
- What about a dirty bomb or a nuclear attack? What would you do?
- What about a chemical or biological agent attack? What would you do?
- What do you do when the news says we have an orange alert? What does that mean?

So, you can see from just these few questions that our civil defense preparedness has not really changed. The general civilian population still has no idea what to do in the event of large scale emergencies. A real large scale emergency like a global invasion or a large meteor impact would be disastrous.

Our response to 9/11 was heroic. All those who participated in the effort are due our nation's gratitude. They were true heroes. As brave as those efforts were, we could have been better prepared.

We should note that the events following 9/11 have taught us a lot more about how to build a military. We must recognize that the opposition is agile and learns quickly how to adapt to our tactics and weaponry. We need well-armed, agile, adaptive fighting units of our own to successfully combat the current foes. We are past the time when we need to have a military based on engaging the monolithic "Bear" called the Soviet Union. We should recognize that a military built to a rigid 25 year plan based on conventional war, whatever that may be, or an all guerilla combat may have little relevance at the time we need it most.

How much smarter and more agile than current and conventional foes will ET be? How much more powerful will his weapons be? We concede that the answer to either question is beyond our knowledge. However, we believe that it is imperative that we as a society begin to develop a plan for responding to the possibility of an attack by ET. Our intent is to call for preparation for a visit from ET as opposed to forecasting an attack. While we are unaware of a pending visit by ET, the probability is greater than zero. We should have a plan even for a friendly visit.

The great Asian Tsunami on December 26, 2004 demanded a different type of response from that of 9/11. Whereas the attacks of 9/11 were localized and killed on the order of two thousand people, the Boxing Day Tsunami, as it is also called, killed in excess of a quarter of a million people in an enormous area around the Indian Ocean. Every bit of the recovery was a responsive effort. There was a total void in preparation for this disaster.

Another major disaster took place in Pakistan – an earthquake. As of November 2, 2005, U.N. officials had put the death toll from Pakistan's October 8 earthquake at 73,000 with at least another 69,000 people injured. The figures, from the U.N. Office of the Coordination of Humanitarian Affairs, covered only casualties in Pakistan. Other parts of South Asia were also affected by the 7.6 magnitude quake. There were as many as 3.5 million people still needing medical care. Another 3.2 million remained homeless and vulnerable as winter approached. Those left without shelter and provisions were left in severe life threatening situations facing the harsh weather coming over the next six months. Pakistan and the U.N. are ill equipped for the pending disaster awaiting in the winter of 2005 and early 2006.

The recovery from the earthquake was an entirely responsive effort. Again, there was an absence of preparation. This was true even though there had been quakes in the last hundred years. It is unreasonable to expect to have sufficient medication, tents, blankets stockpiled for a disaster of this magnitude as they could easily perish while stored. However, having a plan in place to obtain and distribute the required commodities would have improved the efficiency of the relief efforts and probably would have saved a considerable number of lives.

These are a few recent examples of disasters that required significant recovery efforts. They include the terrorist bombings of Spanish trains in March of 2004 which killed 191 people, the terrorist bombing of the London subway trains in July of 2005, the Chinese SARS outbreak and the bird flu pandemic, not to mention the various disease epidemics in Africa. These examples teach us that a single plan for all disasters will most often lead to poor results and the possibility of greater damage to property and greater potential loss of life. They teach us that we need a functioning plan for civil defense that is responsive to the results of the disaster at hand. We need a plan that includes a plan for a visit from ET. That plan should cover both friendly and hostile encounters with an unknown ET. It is important to have a plan for a friendly visit to attempt to prevent that friendly visit from turning into a hostile visit.

So where does this leave the general population? Actually, if how we have typically handled large natural disasters and terrorist attacks is any example, the general population will probably not fare well. The reason for this is that there is little to no civil defense preparation for very large scale disasters.

Consider hurricane Katrina in 2005. The large storm impacted the southern U.S. and practically destroyed the region from Mobile, Alabama to well West of New Orleans, Louisiana. Most Americans watched on television how the residents of New Orleans reacted to the disaster, how the local, city, state, and federal government emergency efforts responded.

Everyone had the opportunity to repeatedly view the vivid scenes of people trapped by flood waters in New Orleans after hurricane Katrina in the summer of 2005. Unfortunately, what television concentrated on represented a small sample of the results of that storm's impact. We learned that a well-planned evacuation is critical to saving lives in the face of such a natural disaster. Had Louisiana and New Orleans had a better evacuation plan, many of those trapped in the flooded areas would have been out of the city. Those who were unable to leave would have had a better situation than the one they endured. Those who were able to leave that stayed, that appears to be a large percentage of the people seen on television, made enduring the situation as well as the rescue efforts much more difficult than they had to be.

The residents of New Orleans handled the situation very poorly. With several days of warning there were still hundreds of thousands that did not evacuate. There were a majority of residents who expected that the government (with no real understanding of what part of the government) would take care of them. There was another contingent of the residents that stayed with criminal intentions – indeed there were organized looting gangs.

The city and local government was overwhelmed and decimated. Before the hurricane there were bus evacuations and broadcasts to evacuate. The mayor of New Orleans did ask for state and even federal level support, but the process for getting that support was somewhat misunderstood. The state dragged its feet on asking for federal support and there are speculations that some of the delays were politically motivated.

Whatever the motivation, the official request for federal help did not occur until days after the disaster. It should be noted here that days after the disaster there were still thousands of local residents waiting for help rather than helping themselves. Granted not all of the locals were in positions to help themselves, but many who were did nothing and waited for the government to take care of them.

Finally, when the federal government was given the authority to move, things began to shape up. In fact, many of the available national military assets had been loaded with rescue supplies for days but could not legally move until the proper requests were made.

Once all the legal issues were overcome one of the largest airlift evacuations in history took place – rather flawlessly and very impressively.

The heart of the problem with the hurricane Katrina situation was based on our cultural acceptance that the government will always be there to take care of us. In the event of an alien invasion the government's resources will most likely be taxed beyond its capabilities. It is very unlikely that help would be available even at the level of the hurricane Katrina scenario. Imagine the devastation of a hurricane Katrina in every major city across the country. Government support will be unlikely. Even if just a few simultaneous disasters occurred at once, our Government or any other would be completely overwhelmed. We are without a plan for such a scenario.

This was a disaster of far greater magnitude than the Federal Emergency Management Agency (FEMA) had ever seen or planned for. Yet, we know that the possibility of a more powerful storm exists. We need to develop a plan for more powerful storms and greater devastation. It is likely that ET could produce far greater damage with less warning and in a shorter period of time.

Hurricane Rita presented us with another situation we were unprepared to handle. Evening news video footage showed cars lined up for miles along the Interstate highways leaving Houston. We heard of people finding service stations out of gasoline, food and water. We have to have a more efficient plan for evacuation of a major city.

People streamed out of Washington, D.C. on foot after the airplane attacks of 9/11. Those people sitting in cars lined up on the Interstates around Houston and the ones on foot leaving Washington, D.C. would have been easy targets for terrorists or even easier targets for an airborne ET attack. What would we do to evacuate New York City, Chicago or Los Angeles if the need arose? And could we possibly do it quickly enough? Not likely.

So, the best civil defense in the case of an alien invasion is to be prepared for it. We are not prepared for such a disaster on any level. Even during the Cold War when a large scale nuclear war was possible at any moment, national civil defense preparedness was very poor. The individual must have a plan and act in order to survive.

During the Cold War most American people had no idea where a fallout shelter was located. And it was very unlikely that the general population was prepared for a large blast. There were most certainly some with bomb shelters who were prepared but they were few and far between. Interestingly enough, most of the folks who were prepared were culturally labeled as extremist, fanatics, and survivalists. Most likely, these folks would have taken appropriate action and would have been unaffected by an event like hurricane Katrina.

Now consider various science fiction examples like *Deep Impact*, *War of the Worlds* (2005) and *Independence Day*. In each of these examples there were major traffic jams, pedestrian jams, and they were all walking or running aimlessly with very little strategy as to where to go and what to do. In *Independence Day* the federal government waits until the very last second to accept that the alien presence is possibly an invasion fleet. And major parts of the general population had the same problem – they had somehow convinced themselves that the aliens were our friends.

Assuming we continue our standard political and cultural approach of reacting to events rather than preparing for them you should assume that your level of preparedness is completely up to you. It is unlikely that we will have major state and federal preparations for alien invasion. We know that a large meteor will impact Earth sometime in the next 10,000 years and we have zero preparedness for it. This does not mean that it will be 10,000 years before the next big impact. It means that a big impact could happen tonight, next week, next year, next century, or sometime within the next 10,000 years with equal probability. Since there is debate about when or if an alien invasion will occur, there are probably few if any politicians who would dare sponsor a spending plan for preparation.

Should the general population be encouraged to develop family sized shelters as in the 1950's during the Cold War? This would be difficult to encourage since the threat is from an Extra-Terrestrial source that many do not perceive as a threat. It is possible that the threat of a nuclear attack might be used as a cover story to influence the general population to invest in family shelters. There would also be a need for economic incentive such as a near full cost tax break to families that have them. Otherwise, it would be unlikely that the general public would make such large investments. Most people could not afford the cost of shelters. Also, the design of the shelters that qualify for the tax break would be passed down from a federal organization. The federal organization, which created the shelter design, would have to know the purpose for the shelter in order to design it. This gives a new meaning to the phrase "tax shelter".

Perhaps city or state governments could be given "pork barrel" projects to construct large community shelters. These community shelter projects could be forced to follow federally mandated shelter designs. The design of these shelters should be determined by the same team of experts chosen to create family shelters to withstand an ET attack.

If it were deemed too difficult to create a "tax shelter" or a "community shelter" program, then the ET attack shelters would have to be created by the federal government and under a veil of secrecy. Otherwise, explaining that a national grid of ET attack shelters is being created would panic the public that the government knew something they were not telling. Or, enrage the public for spending hard-earned tax dollars on crazy ideas such as a national grid of ET attack shelters. The reasons for secrecy of such endeavors become apparent. But most likely, there will be no such shelters developed – public or secret.

It is up to you to survive!

You are on your own. The government is not going to come to your rescue, at least not for a good while – probably not ever. The best bet is to look at the nighttime map of the U.S. and get to the darkest parts of that map as quickly as possible the instant an invasion threat appears. Get as far away from military bases and population centers as you can. Also, you must be prepared for the darker side of humanity. Your biggest threat at first might be other humans trying to take your escape capabilities from you. *War of the Worlds* (2005) gave a good example of this. Be prepared to take any and all drastic actions to defend yourself from the bad side of humanity. It might seem that staying separate from other humans as much as possible is the best answer. However, not all of us are capable of the roughest part of survival living by ourselves. If you are, then good luck and that is probably the best answer for you.

For those of us who are not quite so sure about how to rough it, conglomerating into small groups that support each other appears to be a good idea. The moving convoy of recreational vehicles, vans, and mobile homes in *Independence Day* is a good example. The drawback to the conglomeration approach is that if you get a group that is sufficiently large you might attract attention to yourselves from the invaders. Hopefully, the government will be keeping the ETs too busy to worry about a caravan of busses, but it is a possibility to continually consider.

Maintain the capability to gather intelligence as long as possible. Tune into radio and television stations for the Emergency Broadcast System and log on to the Internet as often as possible. It is likely that the Internet will last longer than any other information and communications source since it was designed to survive a nuclear war. One thing to consider though is to use wireless type connections and radio communications sparingly as they are easily intercepted beacons which will give away your location. Any advanced civilization should be able to track the radio signals to you if you use them long enough.

Now there are many aspects to individual and civil survival that we will leave for other sources to address. Some of the more obvious things such as food and water sources, first aid and shelter, are things that are common to most survival situations and can be found in survival literature that is readily available from sources on the Internet including the Federal Emergency Management Agency. In fact, there are old survival manuals that FEMA produced during the Cold War that would be quite useful in most survival situations.

Finally, you must survive. In the event of an alien invasion it is probable that sooner or later you will be the last line of resistance for humanity. Just like the Soviet-Afghanistan War in the 1980s where the Soviet might crushed any organized military efforts. The sparse resistance forces of the Afghan people were never originally a major military operation and were ill equipped. But the Mujahideen (translates as "struggler" or "holy warrior") held the Soviets at bay for a period of time that was long enough and costly enough to the Soviets that they decided the campaign was no longer worth the costs. The Mujahideen effectively defeated a far superior force by surviving, avoiding capture and making a general nuisance of themselves. They employed asymmetric warfare tactics and strategies that the Soviets were ill equipped to handle. If ETs invade, you, the civilian survivors, may have to become humanity's Mujahideen or resistance force. Note that no religious jihad implications are suggested here. We only are suggesting that the surviving populace will have to take up the fight against the invaders as there will likely be no military left to speak of. Using the Mujahideen's successful tactics might prove advantageous.

Global Communications Must Survive!

It would also be useful to have all of humanity's defenses available and not just those of a few allied nations in the event of an alien invasion. It is very unlikely that a Global Plan of this magnitude is feasible due to the mistrust between world governments (this will be discussed in later sections). However, it would not hurt for nations to conceive of plans to cooperate with other nations in the event of an attack. This might include developing multiple redundant communications technologies and procedures between all nations in the world. Of course this global communications system could be developed under the disguise of international business or extinction level event emergency infrastructure (such as asteroid impacts) rather than in case of ET attack. The need for the possible cover story will be discussed further in later sections.

Consider the development of the Internet as a good example of a robust communications system program. The Internet was developed in order to offer a communications system within the United States in the event of a nuclear war. When the Soviets launched Sputnik in 1957 the fear of nuclear war from space sparked major research and development efforts in both military and civilian space and high technology industries. The launch was a technological boom for the United States and for remainder of the globe. After Sputnik was launched, the U.S. military founded the Advanced Projects Research Agency (ARPA which eventually became DARPA) within the Department of Defense (DoD) to establish US lead in science and technology applicable to the military.

ARPA was assigned to research how to utilize their investment in technologies that would give the U.S. an edge in the Cold War. A major issue of the time was widespread communications and control during or after a major nuclear war. ARPA started efforts to develop computers via Command and Control Research (CCR). This came after Paul Baran, of the RAND Corporation wrote a report to the U.S. Air Force that packet switching was a viable approach to maintaining the command and control capability for a counter nuclear attack. Then Dr. J.C.R. Licklider was chosen to head this effort. Licklider came to ARPA from Bolt, Beranek and Newman, (BBN) in Cambridge, MA in October 1962. Dr. Licklider, then of MIT, first proposed a global network of computers earlier in 1962 that he would eventually describe as the "intergalactic network".

Without going into major detail of the history of the Internet, suffice it to say that it was a major success. The world is effectively wired for communications. The Internet has more than 70,000,000 hosted sites available on it and is growing exponentially. The technologies involved in Internet communications have been grown by defense and civilian industries ranging from intelligence community efforts to the adult pornography industry. All of these developments, though, have created a massive entity of nodes and servers and communications lines (both hardwire and wireless) that is extremely robust.

In the event of a major global catastrophe such as a meteor impact or an ET invasion, such robustness in our global communications system will be a requirement. The question remaining is just how robust is the Internet. Could it withstand alien invasion? And would it be available to all of the remaining Earth defense forces that needed it?

Perhaps a new "ARPANET" effort to improve the next generation of Internet systems should be considered in order to ensure a future robust global communications system.

The global communications systems should consist of wired and wireless technologies at many wavelengths and with various encryption techniques. Direct communication between all of the world leaders would prove crucial in developing any orchestrated defensive maneuvers or offensive strikes. Technology development in global Internet, cellular phones, satellite transceivers, and other broadcast technologies should continue and be encouraged. Most likely, the communications issue will probably be addressed by the international business community simply due to the perceived need for more timely data transfer between companies. These enhancements would be critical for both civil and military operations post invasion.

2.3 The Tactics of Asymmetric Warfare – a View from Both Sides

From the Alien's Point of View

If a hostile ET attacks, there is a reasonable probability that he will employ asymmetric warfare tactics to make his offense easier and more effective. At the same time his use of asymmetry will make our defense significantly more difficult. The hostile ET may have many degrees of freedom available to him in his application of asymmetry. Eight potential degrees of freedom for asymmetric warfare are discussed here.

Asymmetric warfare is a serious force multiplier. Use of asymmetric warfare tactics can allow a small, technologically challenged group to successfully compete with a force of superior size and weaponry for a short period of time. The most significant force multiplier is gained by the application of asymmetric tactics is when a larger or technologically superior force takes advantage of them. As previously stated, ET will probably be technologically superior to Earthlings.

ET could exert COMMUNICATIONS CONTROL. ET's first action could be to assume command of our communications network as was shown in *Independence Day*. We should anticipate that ET will possess the technology to collect COMINT against our communications at will. This capability would provide him insight to our battle planning allowing him to determine the most effective attack scenario. It follows logically, that he will be skilled in ELINT as well.

The ability to perform precise geolocation on our communication signals will probably be in his tool kit. He may be able to jam our communications. He may even be able to take control of our communications networks. He would destroy the timeliness of our electronic communications. We could be forced to return to Revolutionary War communications using runners and paper documents. Using such techniques would be far less timely than the electronic systems our military is accustomed to using. Our creativity could be severely challenged. Having to use such antiquated technology would disrupt and tax our political system. If ET can control communications in the battle space, the disruptive damage to our political structure and its associated psychological impact could possibly be more damaging than physical damage he might impart.

A second degree of freedom in asymmetrical warfare to consider is AIR SUPERIORITY. ET almost certainly will have propulsion technology that is technologically superior to our own. Having more advanced propulsion technology is a necessary, but probability insufficient, capability to be able to travel across a large portion of the galaxy/universe to Earth to wage his war. If he has more advanced propulsion technology available to him, by extension he may have the potential of using one of the favorite tactics of the United States military. He could be capable of establishing air superiority in the battle space. Just as has been true of the United States and the Allied Coalition in Afghanistan and Iraq, gaining air superiority could allow ET to strike targets belonging to earth-bound defenses at will. A good science fiction example of this is the Posleen invaders in John Ringo's *Legacy of the Aldenata* books. The Posleen were able to shoot down any flying craft with ease, forcing the Earth defenses into a one-sided ground based defensive posture. Gaining air superiority would be a major step toward successful implementation of asymmetrical warfare.

CAMOUFLAGE may be the ultimate feature in asymmetric warfare. Camouflage makes it difficult for the enemy to see the camouflage-clad warrior. Camouflage effectively allows the user to conceal himself in plain view. A camouflage-clad ET force could be upon us before we recognized him. He might come wearing camouflage in the form of a Romulan-like cloaking device as seen in *Star Trek*. Another good science fiction example is the camouflage used by the Predators in the movies *Predator*, *Predator II*, and *Alien Vs. Predator*. In such a case the ET camouflage is also a type of armor and is most definitely a force multiplier. A modern warfare example is the technique that the U.S. employs by "owning the night". The U.S. has technology that enables them to operate at night as freely as daytime while its enemies in many cases have no means of detecting soldiers at night. Of course, the more advanced enemies implement infrared systems to minimize the camouflage of darkness. Another good example would be stealth technologies.

The fourth asymmetric warfare degree of freedom available to ET is the ability to adapt his operational focus in order to CONTROL THE NATURE AND TEMPO OF THE CONFLICT. He could shape the battle by selecting targets he attacks. He may have the ability to dictate the places where battles take place. If he can control targets and locations, he would almost certainly be able to select the times of conflict. By exercising those prerogatives, he would be able to avoid engaging us on our terms. His ability to control the nature and tempo of conflict would be another significant force multiplier for him. He would have selected battle features in terms of time, place and targets that would prove most advantageous to him; those features could consequently be most deleterious to our welfare.

An ET who has evaluated the United States' approach to fighting wars would be quite familiar with the fact that logistics are at the heart of our war machine. As a result, ET would probably want to attack our network of roads and bridges around selected military installations such as Camp Lejeune and Fort Benning. He would want to act in a fashion that would isolate the troops in such facilities in order to prevent those troops from mounting a defensive against him. This tactic is standard military doctrine and there is no reason to believe that ETs would not implement similar doctrine in making war.

There is a high probability that ET could impose asymmetry in ORDER OF BATTLE. While picking the targets, the time of battle and the location of battle, he has the option of selecting to use his small, low relatively value assets to attack our large, high value assets. The use of this tactic would either serve as a successful psychologically demoralizing event or it would incite us to a more determined retaliation. The outcome would depend on how affective the small alien assets are in taking out Earth's large assets.

ET is going to be uninformed with respect to the treaties and accords we use as RULES OF ENGAGEMENT when conducting war. At last inspection, ET was absent from the roster of NATO members. And they certainly have yet to be invited to join the ranks of the United Nations. Would he join if invited? He is outside our global political and diplomatic channels. One should expect that he will chose to stay outside those arenas. ET may have studied our gentlemanly rules for conducting warfare such as the Geneva Convention. However, as a nonparticipant in their development, he is free from those rules by which gentlemen try to make the horrible act of war at least somewhat civilized. In truth, war is ugly when we are unable to put a name or face on other combatants.

Through our conventions and treaties, we have tried to make war civilized when combatants are face to face with the ability to recognize each other and especially when one clearly has the upper hand. We believe that it is alright to kill the nameless and faceless, but that we must respect the civil rights of the captured. ET may believe that war is brutal and cruel. He may believe that it is equally suitable to kill those under his control just as it is to brutalize and kill those he is unable to distinguish. He may act in total and complete opposition to our gentlemanly rules of engagement. He may even find that he can use our rules against us. Since we espouse that these rules are a part of our values, we are bound by them. After all, it has been said that values are so integral to our very being that we rarely, if ever, change them. We would be constrained by our rules of engagement even as ET is free to behave in a fashion consistent with his values.

It would be inappropriate for us to criticize him for adhering to his values just as it would be inappropriate for him to criticize us for following our own. Then again, ET might simply consider us pests and plan to exterminate us. Do we consider the rules of engagement that warring fire ant colonies do when eradicating fire ant hills from our lawns?

ET might be able to establish INFORMATION SUPERIORITY as well as air superiority. If so we would have trouble seeing and recognizing him while he would understand and recognize our information networks. He might be able to clearly recognize our communications devices and methodologies while hiding his own from us. He would be able to see our communications infrastructure so that he would have the ability to recognize our strengths and weaknesses. Information superiority has the chance to be more important to the outcome of the battle than the actual physical capabilities.

Information superiority certainly multiplies physical capabilities and weaponry. As part of his information superiority, he is likely to have better intelligence, surveillance and reconnaissance systems and capabilities than we do. His scenario might include selective jamming of communications nodes to effect disorder in our defensive tactics through lack of communication between command levels. He would be able to map our radars and strike them to take them out, jam them, or stay out of their field of regard. Employing either tactic would provide him a significant, perhaps insurmountable, advantage in the battle space. All the while he would be taking advantage of our weaknesses and avoiding our strengths and simultaneously shielding his own vulnerabilities from us. By having such capabilities, he would appear to be elusive and stealthy.

ET will have to be TECHNOLOGICALLY ENABLED AND ASTUTE to reach Earth. We must expect that ET will show up as a result of a plan as opposed to stumbling upon us. He will have to be technologically advanced and smart to get here unless through accident. He will have to be well educated at a minimum at the highest levels of his society. He may have minions or unskilled individuals as the shock troops but he must have some very smart personnel planning the encounter. We should anticipate that even the masses will be as well or better educated on the average than our average citizen.

ET is probably an adaptive species. Remember than he was savvy enough to get here. He is going to have access to military technology. He may also have access to commercial capabilities. In order to have informational superiority, he must have both commercial and military technologies as we define them at his command. ET will understand the value of intelligence in waging war or he will have to have an overwhelming set of weapons available to him in order to be successful. It is reasonable to assign both capabilities to a species that can make the journey ET will have to make to get here – to invade us.

From Humanity's Point of View

The attacking ET is probably going to have superior military weapons. It is possible that he will have more weapons than do we. It is possible; however, that is an unlikely scenario. He would have to transport tremendous additional mass across a huge distance to have more weapons than do we. That additional mass will mean that he needs larger space-travel vehicles or more space-travel vehicles. He would also have to use up a greater fraction of his domestic resources or those of a subjugated society to make such weapons. Even if he has a far smaller number of weapons, he will quite possibly have us out gunned in terms of destructive fire power.

The prospect of being faced with a situation as described lead us to consider a couple of 2005 like analogies. ET would be in a posture of military superiority in terms of weapon sophistication and power in the Global War on Terrorism (GWOT) similar to the posture we enjoy in the fight against al-Qaeda. Humanity will be in a role analogous to that of al-Qaeda in the GWOT. When we are faced with employing tactics similar to those commonly used by al-Qaeda, will we consider it terrorism or will we consider it survival warfare?

The question before us becomes "How do we survive in the face of such a devastating foe?" One viable option is that we turn to asymmetric tactics of our own. ET will take advantage of the degrees of freedom in asymmetry which best serve him as force multipliers. We will need to adopt those asymmetric tactics which best serve as a force multiplier for us in our efforts to survive the attack by ET.

One well known and very successful asymmetric set of tactics that might be used served the fledgling colonists in the Revolution against Great Britain. We could resort to "guerilla" warfare. We could dust off the "Swamp Fox" approach and take advantage of our knowledge of and intimacy with the planet. We could use small, agile low-value assets to attack larger, high-value assets belonging to ET. We would need to strike quickly from ambush and move away rapidly. We could use improvised explosive devices as the insurgents in Iraq are doing in 2005. Our own set of Osama bin Ladens would emerge from the general population over time. They might include current elected officials; however, it is most likely that they will arise from the populace at large. In fact, as discussed previously, we should study and learn from the tactics that Osama and the Mujahideen used in the Soviet-Afghanistan war.

We might resort to the pseudo lawlessness of the "Old West". Like the "Old West," every man and boy would have a gun – at least one gun. Each one would be expected to use it to attempt to be a disruptive force against ET's imperialism. A negative of this approach is that we would be sending farmers and children as overwhelming underdogs to fight well-trained soldiers with superior weapons.

One must ponder the impact of such tactics on ET. Remember, ET will probably have a different set of values. ET may be like the insurgents of 2005 and have values that are in direct opposition to those of the "civilized" nations including the United States, Great Britain and the other members of the coalition in the GWOT. He may be absolutely unconcerned with collateral damage to the "innocent bystander." It is possible that he will place a value of zero on noncombatants or that he sees killing those same people as a benefit in accomplishing his mission. He may retaliate by killing everyone within his range. We need to carefully consider his response to an asymmetric tactic before we commit to it and put it in play. But if the ET invaders have no regard for humanity then any and all tactics would have to be employed for any hope of survival.

We need to drag the fight out as long as possible. The longer we can resist ET, perhaps the better our odds of survival become. We must resupply our warrior population in order to make extending the fight work to our advantage. We could need to revert to historical division of labor. Women may have to go back to pre-woman's suffrage roles in the society and give up all their gains to enable society. We would have to reproduce and reproduce rapidly.

To that end, every female with childbearing capabilities would have to be pregnant as often as possible. Each would have to essentially stay pregnant in order to overcome the mortality rate resulting from such a lopsided war. Each woman would need to take fertility pills in an attempt to multiply the offspring from each pregnancy. Would we be willing to forego our social standards related to fornication, monogamy, marriage and adultery, age of consent, and the age which is commonly considered too old for childbirth in order to preserve our species? That might be a necessary tactic in order to survive as the size of the male population is diminished in the war for survival.

Would the human race be willing to unite to fight the common foe? Or would we still be divided and fragmented along racial, religious and ideological lines? Exhibiting these types of divisions would almost certainly reduce the probably of survival of the human race. The alien invasion scenario was often considered by President Ronald Reagan as his "ultimate fantasy" in which all of humanity would come together as one race to defeat a common foe from outer space. This ultimate fantasy has major ramifications on what we now consider cultural taboos. It is interesting to ponder what a surviving culture might be in this circumstance.

2.4 Earth Defense Technologies and Tactics

Whether national or global concerted defensive efforts are utilized, all of the talents and technologies of the conventional military as well as any unconventional military should be implemented. If the ET military does not destroy our planet outright from orbit and if their intent is to occupy Earth, ET ground forces will be required. All aspects of our conventional ground forces would be deployed. New military ground force strategies for large instantaneous enemy ground forces placement needs to be addressed.

Modern warfare demonstrations and skirmishes such as the Gulf War, suggests that it takes several days to months for American forces to be deployed and prepared to make war (at an appreciable scale) on an enemy force. This is much too slow if the enemy suddenly appears out of warp space into orbit and beams or drops millions of ET batteloids, hovertanks, and starfighters across the globe or even ravenous hordes of predators such as the alien Posleen in John Ringo's *Legacy of the Aldenata* novels. We would not have enough bullets to shoot millions of aliens as they ravaged the planet as was the case in the Ringo books. We will assume that the ETs have such things as battleoids, hovertanks, and starfighters (and other common types of mecha described in anime, books, and film that are beyond our current technological capabilities) in order to make the discussion make sense. It seems logical that ground, armored, and air forces are logical components of any military. However, there could also be alien concepts used that we have yet to imagine such as individual transportation devices, gravity modification beams, shields, nanotechnology, mind control, and who knows what else. We will have to be able to adapt defensive techniques for those alien concepts in a hurry. But, we can at least be prepared for military concepts that we do understand.

One could assume that the deployment of the alien ground forces would occur only after the ET Mother ship or armada has burned down all visible military bases from orbit with their high-powered Directed Energy weapons. Therefore, our global ground forces would most likely be decimated. Even if they were still in tact due to being hidden deep underground, how could they be deployed in time to defend against the rapid ET deployment?

New concepts and strategies for ground force rapid projection or continuous global deployment should be investigated. There should also be a second and third tier of troops that were well hidden and protected deep underground that could be implemented more cautiously and surgically that will enable us to prolong our defensive posture. This is the reserve concept that was so successful for Hitler in World War II. Since Hitler maintained a major reserve army in Germany, he was able to prolong the war much longer than it might have lasted without those backup resources.

Air power would also prove to be very important during the early phases of the ET onslaught. Modern day military fighter and bomber aircraft should be deployed globally and in great numbers. The Gulf War shows us that air superiority is a must in any war. The problem is that most likely the ET starfighters and hovertanks will totally out match and over power our aircraft. New and improved aircraft technology must be pushed to the utmost edge of a new more complex state-of-the-art. It is likely that our modern aircraft technology will not last long against the ET air power. We must stop designing for the inferior threats we know or can easily postulate and start incorporating design and performance parameters for threats far superior to those we know. We need to bring together a team of free thinkers who eschew "Normal Science." This group needs to be the ultimate "think tank" comprised of the smartest people in the United States and possibly close allied nations. It will be necessary to turn over the leadership and add new membership on a routine basis to maintain creativity. This approach needs to be applied to all types of weapons design, not just aircraft. However, if we begin to design fighter craft now with the purpose in mind of defending against an infinitely more powerful enemy rather than equal Earthly enemies, it is possible that we could develop air defenses that would at least last longer against the ET onslaught and could dominate enemy forces on Earth.

Potential Point of Vulnerability of Attacking ET's

When we consider defensive weapons against ET's crafts, we hope that he has not found a method of minimizing the effects of our gravity as interpreted through our understanding of physics. Keeping a craft on orbit and maintaining agility requires us to make it as light as possible. We also want a powerful propulsion system to allow maximum agility. If this is the case for ET then he is possibly going to need shields with selective reinforcement in vulnerable areas; these might add difficulty to remaining agile. If ET is constrained by metallurgy similar to our own, he will have vulnerable points in his structure in order to make the craft flyable. His craft will probably not be as thermally vulnerable as ours or he would have most likely suffered damage to a fairly high percentage of his fleet upon entering our atmosphere. Unless he has developed a shield capability, his craft would probably be vulnerable to high speed projectiles or to explosives.

The fact that he got here probably tells us that he has propulsion technology beyond our current level of knowledge. Therefore, he is almost certainly capable of greater agility per pound for his craft than we are. There are three possible alternatives to his having a more advanced propulsion system than do we. First, his species lives incredibly long lifetimes. The second option is that he consumes virtually no food and has virtually no waste products. He would almost have to take energy very efficiently from the environment or perhaps even from the vacuum energy of spacetime itself as mentioned in Travis S. Taylor's *Warp Speed* and *The Quantum Connection.* Third, his waste products are recyclable to produce new food stuffs and required environmental gases. Some combination of the three is probably required unless it is a virtual day trip for him or he is traveling at incredibly high relativistic speeds. We've also briefly mentioned "shields". Before we go further we should perhaps discuss this concept a bit.

Combating Shields

It is likely that in order to survive the rigors of interstellar space travel the alien spacecraft are protected by either a material that has amazing stress and strain properties or by some other means. The typical science fiction description is "shields". In some cases these shields are electromagnetic in nature, they use gravitational means, quantum flux properties, or some other science fiction "magic wand" that creates a "force field".

It is unlikely that we will determine the nature of the alien shields, but in order to defend against them we must be aware of them and have at least a minimal tactic to confront them. For the force field type of shields like those of *Star Trek, Independence Day, War of the Worlds* (2005 version), and many others, science fiction offers us a possible tactic.

The shields are a bubble or a field around the ship. The tactic offered in all of the above mentioned science fiction examples is to have some means of getting inside the shields and then using a destructive weapon appropriately sized for the vehicle. In the case of a large Mothership scenario like *Independence Day* or the episode of *Star Trek:TNG* where Picard was taken capture by the Borg the solution was to fly a vehicle through the shields and then perform actions once inside. In *Independence Day* the heroes were fortunate enough to have an alien ship with codes that enabled it to pass into the interior of the Mothership. This is an unlikelihood at the onset of the invasion.

In *Star Trek:TNG* the crew of the Enterprise had monitored the Borg shield modulations and adjusted their own shields to allow them to pass through, but only after being weakened by an attack. Since humanity has yet to invent shield technology this scenario is also unlikely.

However, the tactic is sound for a longer term solution. A detailed analysis of the shields might offer insight into how to penetrate them – if only with a single explosive device.

Another thing to consider is that for the aliens to fire their weapons through the shields there must be a technology that allows for that. One of the most important first actions of Earth forces must be to have detailed sensor observation of the alien weapons and shields interacting. Detailed sensor data from Earth technology impacting the alien shields would also be useful.

A second scenario is that the aliens are abducting, capturing, or storing people or equipment. The scenario in *War of the Worlds* (2005) where the alien craft were capturing humans and pulling them inside for whatever alien motive is an ideal scenario for a valid military tactic. A possible tactic would be to have specialists allow themselves to be captured. The specialists are armed with various highly destructive weapons and once abducted and inside the shields they unleash the weapons on the craft.

Another tactic that could be used if a more gruesome and dire situation exists would be that of the suicide bomber. Nonspecialists could sacrifice themselves to the aliens. If they are strapped with explosives and the aliens abduct them through their shields, the suicide bombers can then release the boobytraps.

Unfortunately, either the specialist or suicide attack will only work so many times before the aliens become wise to it. An orchestrated effort of many specialists attacking with this tactic at once would yield the best results.

Again, detailed analysis of the interaction of the alien shields with all types of impact needs to be conducted assuming that time and the scenario allows. Any data might offer a means to defeat the shields.

Considering energy levels required to overcome the shields suggest that it is unlikely the shields can be overwhelmed by Earth forces. Consider an alien craft that can reach 1000 times the speed of light. A single particle that weighs a billionth of a kilogram would impact the vehicle (during interstellar flight) with an energy of over 10^{13} Joules. This is on the order of a small nuclear explosion. It is quite likely that many such particles will be encountered during an interstellar spaceflight. So, it is not an amazing leap of faith to suggest that such interstellar craft must be capable of protecting themselves from energy levels many times that of a nuclear device. How they would do this is beyond our current understanding of physics.

Some possibilities of how the alien shields might handle such energies are, however, within the bounds of modern physics speculation. One speculation is that the shield implements some phenomena of physics that simply does not allow for penetration at any energy. An example of this is the theorized Alcubierre warp theory. According to several publications within the current general relativity literature, "faster than light" travel via a spacetime warp is possible (although the energy levels required are exotic and large). Some further analysis by the community also suggests that the warp bubble would not allow penetration of information/energy through it while at warp speeds because of possible causality violations. Perhaps a concept implementing this physical phenomenon would allow for shields with no upper limit on the energy they could repel. A science fiction example and discussion of these types of shields can be found in *Warp Speed* and *The Quantum Connection* by Travis S. Taylor.

This section has been a brief discussion of shields and a few possible tactics for fighting against them. More speculation, analysis, and brainstorming needs to be conducted on a large scale to understand the possibilities of such technologies and how they would impact the outcome of combat situations. It is obvious that any special protection would be a major force multiplier and would prove difficult to defeat. A modern analogy of this situation are the American operations in Afghanistan and Iraq (2001-2005) where the U.S. troops have state of the art ceramic armor that protects them from most firearms caliber weapons. The opposition force does not have such armor and therefore they are forced to use different tactics and strategies to face the U.S. forces. The main tactic is to use improvised explosive devices (IEDs) rather than an aimed and area fire attack. Against an alien invasion we would be forced to similar tactics with much more powerful IEDs.

Weapons of Mass Destruction – the Chemical, Biological, & Nuclear Options

Typical American defense strategy is that once our conventional military has either been fought to a standstill or defeated, other technologies such as chemical, biological, and nuclear (CBN) weapons of mass destruction (WMDs) may prove useful. We always wait until the very last possible moment to use WMDs. Against ourselves this makes sense. We do not want to destroy our habitat if we can avoid it. Against ETs, on the other hand, waiting to the last minute to use WMDs might be the wrong strategy. We need to give serious consideration as to when would be the proper time to use WMDs during the ET attack. A detailed study needs to be undertaken.

One can consider the scenarios put forth by multiple different science fiction novels, movies, and television series where CBN-WMDs played a major role in the final battle or last stand against the alien offensive. How much of that approach is science fiction and how much of it could be fact?

Without knowing something about the anatomy and physiology of the ETs, it would be difficult to predict which chemical or biological agents would have any effect on them. Therefore, chemical and biological WMDs can only be used as a second tier or later weapon. In other words, CB-WMDs are only useful once intelligence about the ET attackers is gathered and analyzed. Then a possible CB-WMD can be developed and deployed. Rapid deployment in the first tier attack would most likely be ineffective and therefore out of the question.

Depending on the power classification of the ET war machine, nuclear WMDs (N-WMDs) might be useful. Against ETs of level α_1 or α_2 (as shown described in Table 1.1), N-WMDs might be effective. It is likely that the technology level of these ETs would not be great enough to withstand a direct hit from multiple megaton nuclear blasts. However, ETs with power levels above α_2 may be manipulating much more energy than involved in such a blast on a regular basis in order to power their interstellar propulsion capabilities. N-WMDs as we know it would probably be very ineffective against the greater than α_2 ETs. The initial atomic bomb test at the Trinity Test Site and the subsequent bombs dropped on Japan during World War II each released about 1×10^{14} Joules of energy.

Assume that an α_2 ET army moved into Earthspace. This would imply that millions of kilograms of mass due to the soldiers, starfighters, hovertanks, and battleoids, as well as other infrastructure in the ET mothership or fleet were propelled at hundreds of times the speed of light for a significant distance. We have no technique for determining the kinetic energy of such a vehicle, but assume the kinetic energy equation is valid (this may or may not be a very conservative estimate; it is uncertain if such an equation for warp speeds is legitimate). A more detailed discussion of General Relativity would be required but is outside the purposes of this text. Then

$$E = \frac{1}{2}mv^2 = \frac{1}{2}\left(1x10^6\,kg\right)\left(100\left(3x10^8\right)\right)^2 = 4.5x10^{26}\,Joules\ ! \tag{2.1}$$

The kinetic energy of such a spacecraft moving at 100c is many orders of magnitude (more than 12) greater than a nuclear explosive. Note that if special relativistic effects are considered the energy is even greater. Even the largest nuclear devices available on Earth today are many many orders of magnitude smaller in energy than the kinetic energy of the ET space fleet.

The largest bomb in human history was the Russian *Tsar Bomb*, which was about 6,500 times larger than its World War II ancestors. The Mk/B-41 is the American version of the *Tsar Bomb* and is shown in figure 2.9. So, it would have produced about $6.5x10^{17}$ Joules. Therefore, the biggest bomb humans have to offer is still nine orders of magnitude smaller than the minimum energy requirements of an α_2 ET army.

It might be however, that if the bomb could be detonated inside such an enemy's defensive armor or shielding technology, then serious damage might be inflicted upon them. Of course, there is no way of knowing how to do such a thing. Placing a bomb inside an ET's defensive perimeter is definitely a second or third tier action. We must survive long enough to gather much intelligence on the ET defenses in order to take such actions.

Perhaps a modified first tier would be beneficial. We could implement National Missile Defense (NMD) technologies like Brilliant Pebbles. Brilliant Pebbles is to be a constellation or swarm of killer satellites that are placed in orbit and remain dormant until there is a space born target to pursue, impact, and destroy. Presently, Brilliant Pebbles would be designed with conventional warheads for the purpose of destroying ballistic missiles in the post boost (or midcourse) phase of their flight trajectory. The anti-ET Brilliant Pebbles could perhaps be a satellite swarm of autonomous Mk/B-41s. In the event of the ET armada showing up, we would at least already have in place hundreds or thousands of the most destructive devices we could muster in orbit and ready to pursue and destroy.

Of course, this concept violates the "no nuclear weapons in orbit" part of the Outer Space Treaty of 1967. And they would most definitely need to be a well-guarded secret to protect from accidents and terrorist incidents as well as sparking a new Arms Race. However, it is a good plausible first line of defense.

Public domain image from http://nuketesting.enviroweb.org

The **Mk/B-41** was the American version of the *Tsar Bomb* and the highest yield nuclear weapon ever deployed by the U.S. It was also the only three-stage Ulam-Teller thermonuclear weapon ever developed by the U.S. Note that the mass of the bomb could easily be lifted by a Delta IV expendable launch vehicle to elliptical orbits with an apoapsis as high as 35,000 miles.

Yield	25 Megatons
Weight	10,670 lb
Length	12 ft. 4 in (148 in)
Diameter (body)	52 in
Diameter (tail fin)	74 in
Number Manufactured	About 500
Manufactured	September 1960 to June 1962
Retired	November 1963 to July 1976

Figure 2.9 The American version of the Tsar Bomb the Mk/B-41

Another N-WMD that might be useful during the ground occupation phase of an ET attack would be the miniature nuclear bomb called the W-54 warhead, which was designed to fire from the Davy Crockett launcher (see Figure 2.10). It was deployed by the United States during the Cold War and was to be used on advancing Soviet troops if the need were to arise. The weapon would be the ultimate one shot artillery! The W-54 weighed about 25 kg, could be launched from the man-portable or Jeep portable Davy Crockett launcher, and would cause an explosion about 200 times smaller than the Hiroshima bomb or about 5×10^{11} Joules.

These smaller nuclear bombs might prove useful as the active warhead on the antistarfighter, antihovertank, and antibattleoid missiles that we should equip our fighter aircraft and ground vehicles with. It is possible that such compact but high yield explosives may affect the smaller ET craft's armor/shields. These antistarfighter missiles most closely resemble the AIM-26A Falcon class of air-to-air missiles, some of which were tipped with the W-54 warhead of course modern sensor and missile designs might be more effective. The Davy Crockett launchers could also be used on ground armored vehicles to combat the ET hovertanks and battleoids.

Knowledge of the build-up of such weapons might change global politics. Therefore, extreme secrecy of such preparation efforts would probably be warranted. Also and unfortunately, it is unlikely that all of the world's governments would believe that we were building these weapons for their own good. Most definitely if our non-allied friends found out about our military build–up, a new Arms Race would ensue. Utmost secrecy would be a necessity and may prove to be a requirement for the survival of the human race. On the other hand, a new Arms Race might actually improve humanity's chance of survival against an ET invasion – the ramification of humanity surviving itself due another Arms Race is another question to be considered.

At any rate, small highly volatile weapons systems like the W-54 tipped missiles would be an absolute must for defense against an alien ground and air force. It is extremely unlikely that any of our conventional surface-to-air and air-to-air weapons would have any success at all in stopping the alien attack systems.

What about Chemical and Biological?

It is so unlikely that we would know enough about our alien invaders to implement a chemical or biological weapon against them initially that there is little we can do in this arena. Perhaps after the initial attack and we understand more about their physiology these weapons might become a factor. For now, it is our opinion that there is little to be done until more intelligence on the alien is gathered. Perhaps we are incorrect and this area of WMD should be investigated further?

Figure 2.10 The American Davy Crocket Launcher and W-54 tactical nuclear warhead might be a potential anti-ET vehicle weapon

Directed Energy Weapons – Lasers aren't Just for Science Fiction

There are also various efforts presently to develop fairly high energy Directed Energy or Beam Weapons (DEWs). Most of these weapons produce an infrared laser beam either through chemical interactions of hydrogen (or deuterium) and fluorine or oxygen and iodine. There are other concepts being investigated but the chemical lasers seem to be most likely to be deployed in the near future. Test lasers of this type like the Mid-Infrared Advanced Chemical Laser (MIRACL) shown in Figure 2.10, which is the first megawatt-class, continuous wave, chemical laser built in the free world might be useful in defense against the ET attack.

MIRACL is a deuterium fluoride (DF) chemical laser, which produces infrared laser photons in the range between 3.6 and 4.0 microns wavelength. It first lased in 1980 and is located at the High Energy Laser Systems Test Facility (HELSTF) in White Sands, New Mexico. To date, it is the only US laser, which has attained average powers in the megawatt range.

Assuming the MIRACL laser is the largest available DEW and if it can fire continuously for about sixty seconds before it needs refueling, then it can deliver a million watts of power for sixty seconds. In other words it will deliver a beam of energy to target of sixty million Joules or $6x10^7$ Joules. High explosives on missiles might have more energy than that, which can be delivered to a target, but the DEW's energy beam travels at the speed of light. So, DEWs have miniscule lag time between firing and impact at combat theatre ranges. Such fast delivery might be required in order to target and hit the ET starfighters and hovertanks. Advanced controls and ability to withstand high g maneuvers might make it difficult to hit an ET craft with conventional missiles. The DEW offers an alternative that travels at light speed rather than hypersonic speeds. More DEW research is needed. Much higher energies, longer firing times ("deeper magazines"), and more numbers of them need to be produced. A network of such devices should be deployed, if not globally, at least nationally.

Steps in the right direction toward this network of deployed laser weapons systems are the Tactical High Energy Laser (THEL) and the Mobile Tactical High Energy Laser (MTHEL) development programs. The THEL demonstration hardware has successfully shot down over 25 Katyusha missiles and even a mortar round. The THEL and MTHEL systems are shown in Figures 2.11 and 2.12. Perhaps the continued development of these systems will enable anti-ET weapons systems that would prove useful. However, the weakness with these systems that should be kept in mind is the fact that present energy levels are fairly low when compared to nuclear alternatives. But, the future of HEL systems could possibly lead to higher energy breakthroughs with different technologies such as solid state laser advancements.

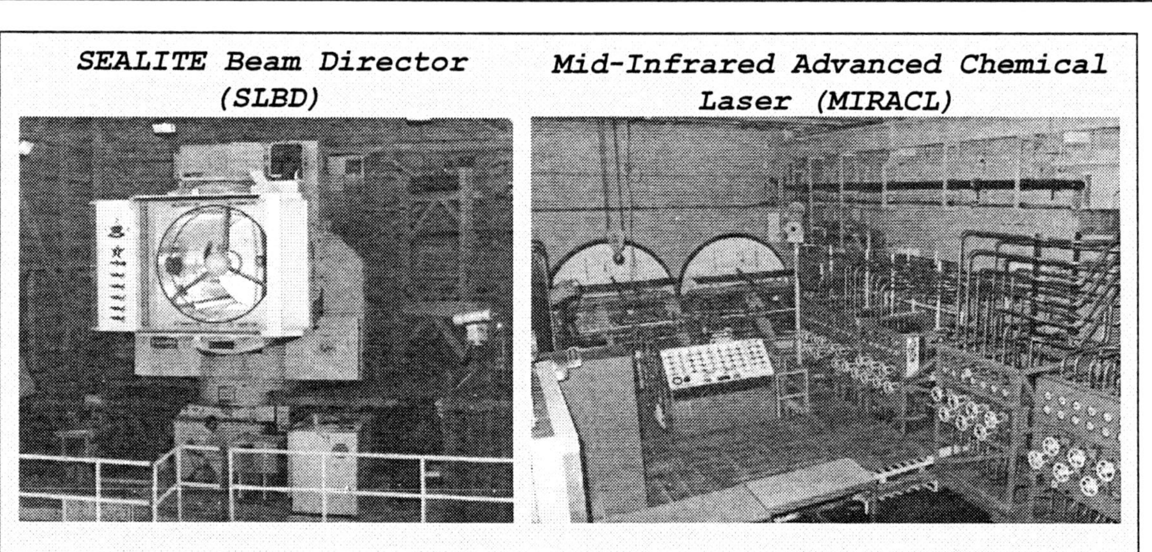

Capabilities

- Megawatt-class variable power
- Pseudo-continuous-wave mid-infrared (~3.8 microns)
- ~70 seconds maximum lase duration.
- Reliable operation demonstrated in more than 150 lasing tests

MIRACL uses the ***SLBD*** to steer the laser beam

Credit: U.S. Army Space & Missile Defense Command

Figure 2.11 The MIRACL chemical laser testbed

Figure 2.12 The Tactical High Energy Laser (THEL). Credit: U.S. Army Space & Missile Defense Command

Figure 2.13 Artist concept of the Mobile Tactical High Energy Laser (MTHEL) system. Credit: U.S. Army Space & Missile Defense Command

Electronic Warfare

Most electronic warfare is theorized as being implemented through Directed Energy Weapons (DEWs). Most DEWs include laser, high power radio frequency (HPRF), and particle beam technologies (HPRF technology is frequently called high power microwave (HPM) or RF directed energy). The electronic warfare concept is that typical electronic equipment can be defeated, impaired, or even destroyed by irradiation from directed energy (DE) sources – a good example would be lightning striking a television or a microwave being too close to a computer or solar flares disrupting communications satellites. Degradation effects can manifest as short-term "upsets" in electronic and electromechanical systems and sensors to enduring circuit damage or to complete destruction due to thermal or electrical discharge and overload. Commercial and military systems and subsystems have become increasingly more and more reliant on microelectronics and therefore have become extremely vulnerable to DE radiation. In fact, both commercial and military programs expend great resources studying ways of protecting microelectronics from EM sources.

It is reasonable to assume that a visiting ET has more highly developed electronic suites than we are accustomed to or are capable of building. ET probably has his systems hardening to attack to levels beyond which we currently envision. The US Air Force has been studying high power microwave for some time. Much of this discussion is derived from Air Force reports. However, their efforts have been aimed at conventional Earthling technology; we must raise the technology bar for ET. Raising the bar will also encompass near-term threats from conventional foes.

High power radio frequency (RF) is generated by the interaction of charged matter with appropriately constructed conductors and dielectrics. High power RF weapons include narrow band to ultra wideband RF devices. No matter what wavelength region of operation of the weapon, high power RF weapons have three essential elements: a radiating structure, a cavity that couples RF energy to a relativistic electron beam and a source of pulsed power to drive the beam and cavity. The high energy densities in such systems cause ionization of some portions of the normally solid materials. We will need to develop weapons across the entire RF spectrum to prepare for ET as we are currently ignorant of his capabilities.

According to reports by the National Security Space Architect (NSSA) "Near-term goals for RFE weapons include the development of new HPRF source concepts, such as the interference modulation HPM source concept and frequency agile, broadband klystrons for use in susceptibility testing and in field tests. A mid-term goal is the development of ultrahigh-gain, broadband antennas. Long-term goals include use of chaos theory research results to achieve greater control of RF weapon sources."

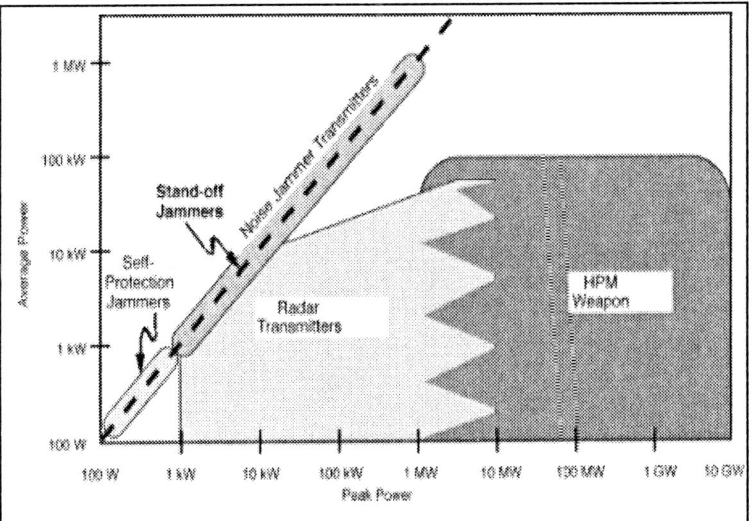

Figure 2.14 RF Weapons Power Categories as Reported by Kirtland Air Force Base, New Mexico, Phillips Laboratory and the National Security Space Architect

Development of smaller, lighter, more fuel efficient high power RF generators would allow us to mobilize them as spaceborne weapons which would probably provide us the greatest deterrent capability for ET. Improvements to reduce size, weight, and power requirements must also be accomplished by enhancements to radiation beam control for space deployment. We must project targets and undertake intensive susceptibility studies to determine the best attack methods. These technical challenges require concentrating technology development efforts on improving modulators, RF sources, and antennas. Figure 2.14 shows power requirements for various types of RF weapons categories. The graph gives us some idea as to the average and peak power requirements for systems that would be effective against manmade capabilities. We can only assume that power requirements for impacting ET systems would be much larger.

The current Air Force Research Labs (AFRL) program is developing measurement and predictive techniques for electromagnetic (EM) effects. New EM hardening technologies are being developed, evaluated, and implemented into operational and new systems, and subsystems. Experiments are planned to be conducted on aircraft such as the F-22 and B-2 that will enable evaluation of commercial GPS receiver units under EM effects. We can learn from these efforts how to expand our technology developments.

Experimental efforts, and consequently weapon development, are typically more timely and cost-effective if guided by scientific modeling and simulation, i.e., virtual prototyping. RF weapon systems can be modeled using Maxwell's equations and either

Figure 2.15 High power RF testbed. Credit: U.S. Air Force Research Laboratory, Kirtland AFB, NM.

the Lorentz Force law (for the kinetics of individual charged particles) or the Navier-Stokes compressible fluid equations (for the dynamics of a conducting gas). The resulting nonlinear system of partial differential equations is analytically intractable, and their numerical solution requires fine-grained resolution. As a result, the computational modeling and simulation of RF devices is expected to be computationally intensive.

Figure 2.15 shows a test bed in the AF High Energy Research and Technology Facility at Kirtland AF Base in New Mexico. Is it possible that RF weapons like the one shown in the Figure could be updated for use against ET technologies? More research should be continued in this arena.

Advanced Spacesuits for Mobile Infantry – We May Need to Fight ET in Space

In the event that a manned force is needed to penetrate an in-space ET vehicle, there are serious life support issues that must be resolved. It is possible that there would be zero to high gravity onboard such a vessel. It is also likely that the atmosphere within the vessel is not suitable for humans. An in-space assault on the ET vessel might also require exposure to the extreme conditions of space. Present day spacesuits are cumbersome, bulky, and fragile. One definitely does not think of a combat soldier when the image of an astronaut in his spacesuit comes to mind.

Figure 2.16 shows the modern day spacesuit or Extravehicular Mobility Unit (EMU) and the state-of-the-art maneuvering system called the Simplified Aid for Extravehicular activity Rescue (SAFER). The SAFER attaches to the back of the EMU and enables motivation through space at distances on the order of a few tens to hundreds of meters.

Also, the EMU is inflated to about one third of an atmosphere. If the outside of the EMU is the vacuum of space and the inside is inflated to a third of an atmosphere, the suit is more like a balloon than a suit. The inflation pressure adds significant resistance to motion of the arms, legs, and hands. It is very difficult for astronauts to conduct the simplest tasks in these mobility restricted EMUs.

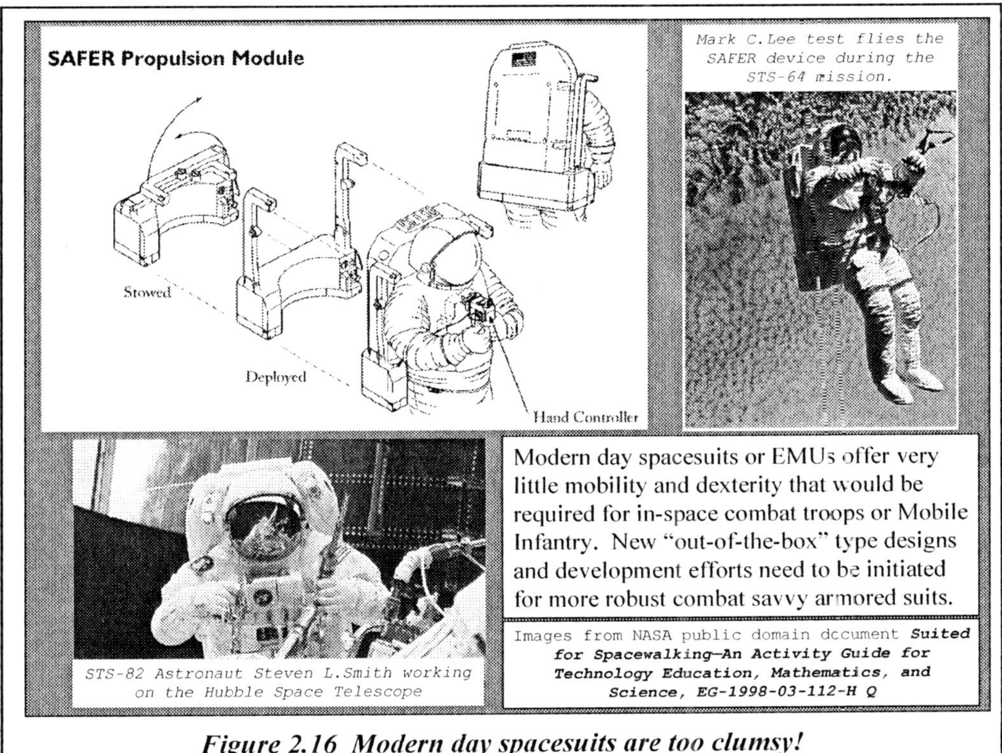

Figure 2.16 Modern day spacesuits are too clumsy!

Combat in such an outfit is absolutely out of the question. Soldiers moving as slowly as modern day astronauts in their EMUs would be sitting ducks and very ineffective mobile infantry. Therefore, an "out-of-the-box" concept for the combat spacesuit for our Mobile Infantry must be designed. We suggest a suit similar to the mechanized armored suits of the Mobile Infantry in *Starship Troopers*. Heinlein's Mobile Infantry wore armored suits that gave mechanically added strength to its occupant as well as armor and firepower. The suits were also heavily equipped with sensors and weapons as well as jump thrusters. These strength enhancing armored suits are also similar to the suit worn by the Green Goblin in the movie *Spiderman*. Other examples are the ACS in Ringo's *Legacy of the Aldenata* books, hovertanks and cyclones and Veritech fighters from Robotech, the Marvel Comic hero *Ironman*, and DC Comics *Steel*.

In order to develop such a mechanically advantaged and armored system new controls and reaction interfaces must be developed. The suit would have to respond to the actions of the body and not only mechanically move the suit in a likewise fashion, but to also move it in a way that does not hurt its occupant. Interfaces that allow direct thought control of electronic actuators and switches would benefit the design of the suit as well. It is conceivable that the suits helmet could be equipped with sensors such as a superconducting quantum interference device (SQUID) that can accurately measure changes in electrochemical brain signals. A processor would then interpret these signals into control commands for the suit. This sensor studded helmet concept is similar to the "thinking cap" used by Veritech fighter pilots in the *Robotech* books. Mention of the SQUIDs has been made in several science fiction movies such as *Strange Days*. DARPA and NASA have funded early efforts to develop human-machine interfaces that would relieve the need for keyboards and mice, but the efforts are in their infancy.

It would be required for our scientific and engineering community to take these science fiction concepts from the realm of fiction to that of reality. Such suits would offer a great advantage not only in the case of an ET attack but also in normal military endeavors and operations. As it currently stands, we would have very little chance of mounting a space attack with infantrymen. The technology is just not available, presently. However, DARPA has a program called the Exoskeleton for Human Performance Augmentation (EHPA) program in which several organizations (universities and industry) have been funded to investigate the technology requirements for the Powered Suit.

Under the EHPA program Oak Ridge National Laboratory completed an effort called the *Exoskeleton for Soldier Enhancement Systems Feasibility Study* whereas many of the technology issues required for a powered armor suit are addressed. The report investigates various control systems, power systems, servo mechanisms, and joint/muscle ranges of motion. Another effort by Applied Motion details a patent of a *Servo Assisted Lower Body Exoskeleton*.

NASA has also had an effort funded through its' NASA Institute for Advanced Concepts (NIAC) office. This effort was titled *Astronaut Bio-Suit for Exploration Class Missions*. This study detailed more of the space environment aspect of the suit whereas the DARPA effort was for terrestrial based soldiers. A true Powered Armor Suit for use against an ET invader would be a hybrid of the two efforts thus allowing for combat in a space environment if need be.

In the 1960s General Electric developed a system called the HardiMan, which was too heavy and the technology of the time could not support the control requirements. The Berkeley Robotics Laboratory has developed a version of a powered assist machine called the Berkeley Exoskeleton or BLEEX. And there are several ongoing efforts in Japan. Figure 2.17 shows some of the results of these efforts and gives a few artists' concept drawings of Powered Armor Suits. Perhaps a serious research effort into Powered Armor would yield a viable weapon system.

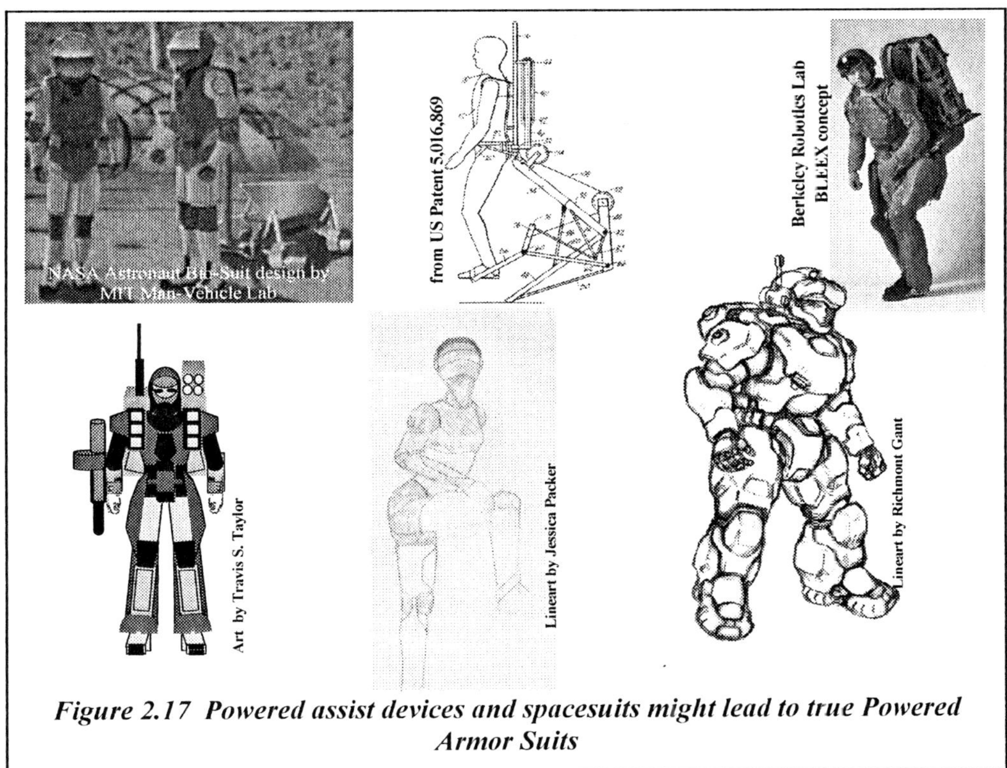

Figure 2.17 Powered assist devices and spacesuits might lead to true Powered Armor Suits

One final note on Powered Armor Suits is that there is some debate on the usefulness of them from a military aspect. Without going into major details of the debate here we suggest that the technology is simply the next extension of armor and environment protection. Troops are already carrying maximum loads and astronauts are severely limited in what tasks they can perform. This is why there are DARPA and NASA programs to develop such systems. Feel free to conduct debates on the viability of these systems as military assets on your own.

Mecha, Armored Configurable Vehicles, and Other Similar Concepts

This section would not be complete if we failed to address a similar science fiction concept – mecha. Wikipedia.com describes Mecha as:

The distinction between smaller mecha and their smaller cousins (and likely progenitors), the powered armor suits, is blurred; according to one definition, a mecha is piloted while a powered armor is worn. Anything large enough to have a cockpit where the pilot is seated is generally considered a mecha.

The H.G. Wells novel *The War of the Worlds* is often credited as the first mention of mecha. In that book the Martian invaders use tripod walkers very similar to what would be considered by the Wikipedia definition as mecha. Other well known science fiction, anime, manga, and video game references are *Robotech, BattleTech, Gundam, Voltron, Robot Wars, Tank Police, BubbleGum Crisis,* walkers in the *Star Wars* universe and countless others. We recommend doing a websearch for mecha for further background. Figure 2.18 shows a typical mecha configuration.

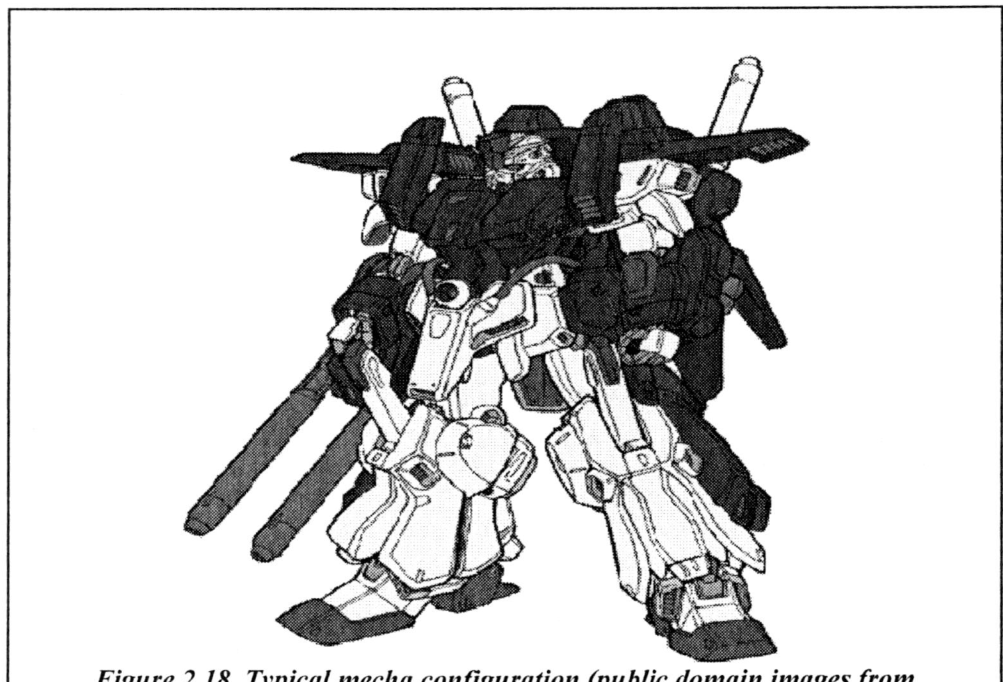

Figure 2.18 Typical mecha configuration (public domain images from HTTP://gundam.anime.net/images/)

On many web blogs, and forums there is a considerable debate about the potential military advantage or disadvantage of these weapon systems. Mecha is typically considered as a science fiction replacement for tanks, fighters, and spacefighters. It is also typically used for logistical support like the walking forklift in *Aliens*.

The typical argument for the mecha is to have them "walk" and maneuver like people. This capability would theoretically give the heavy armored vehicles true "all-terrain" capabilities. On the other side of the argument is that the heavy armored systems would be placing all of their weight on feet rather than large surface area tracks which could mean the heavy mecha would sink into the ground. There is merit to each side of this argument but it is not certain that the mecha would be made of standard materials and perhaps the weight issue could be dealt with using future materials technologies.

There are endless arguments for and against the mecha concept, which is beyond the scope of this text. However, we think that the arguments are based too much on the stereotypical mecha meaning that the arguments are for specific genre or concepts from the various science fiction stories. What we suggest is to study a hybridization of the mecha idea that might lead to an extremely useful weapons system.

Our first suggestion is for the flying, vertical take off and landing (VTOL), armored tank. The usefulness of tanks in the modern military is unquestionable, but the problem with them is rapid force projection. It takes a massive infrastructure and transport system to get the tanks from point A to point B and it takes a considerable amount of time. A tank command in Fort Knox would be sorely challenged to be in New York City in less than a week's time. To get to an overseas arena takes much longer.

Now consider a tank that is as powerful as a modern M1A2 and can fly with VTOL capabilities. Now add to that a flight envelope similar to a modern commercial aircraft. Then add to that concept modern stealth aircraft capabilities. The final design becomes a flying stealth tank with VTOL and a global deployment range of a day or so. If you add hypersonic flight capabilities the tank could be anywhere on Earth in a few hours.

Imagine the capabilities of a squadron of these flying stealth tanks (FSTs). Applied to the war in Iraq, the FSTs could be in Baghdad for a firefight in the morning and in Falujah in a matter of minutes if they were needed. The strategic and tactical implications of FSTs are incredible. FSTs equipped with Davy Crockett nuclear warhead launchers might be just the type of armored vehicles we would need to fight an ET ground invasion force. Consider the FSTs with Davy Crocketts in the *War of the Worlds* scenario. The immediate outcome of the conflict might be different, indeed.

More serious analysis, design studies, and tactical combat simulation need to be conducted with mecha concepts. It is possible that systems that can be reconfigured for flight, or naval applications, or logistics support as well as other combat applications might prove quite useful. A major obstacle for current armored vehicles is river crossing. This would not be a problem for FSTs and other mecha type concepts. More serious thought needs to be given to this and similar concepts.

There are Many Technologies That May be Useful

As a species we have developed many technologies and concepts that might prove useful if threatened by an ET attacker. There is no way that this chapter can encompass all such technologies, in fact it would take more than several textbooks to describe them all. On the other hand, some of the more pertinent technologies are

- Earth-to-Orbit Launch Vehicles
- Lunar Exploration Vehicles
- Micro, Nano, and Pico Satellites
- Unmanned Ground, Sea, and Air Vehicles
- Global Redundant Communications Networks
- Deep Space Observation Technologies (telescopes, radio telescopes, gamma ray telescopes, radars, etc.)
- Advanced Computing Technologies
- And many others yet undiscovered.

Of course, there are many technologies that have been developed which might prove useful in case of an attack. Also, there are many technologies whose development is very deficient. Our in-space propulsion capabilities are minimal. We currently have no capability that will return us to the Moon. For the most part, mankind is stranded to within a few hundred miles above the Earth's surface. Most of the technologies developed for the Apollo program are ancient history and are gone forever since it was not maintained and continuously implemented. There is technology that could be developed that would allow lunar missions, but it would require major undertakings of the planets aerospace industry. If we needed to go to Mars for example, there is very little hope of doing this in a time period of less than a decade of development and testing. It would probably take five years or more to get back to the moon.

If a Near Earth Object (NEO) were to threaten us, such as in the movies *Deep Impact* and *Armageddon*, there is no hope of developing a spacecraft in time to defend the planet. The NEO scenario is as likely if not imminent (just a matter of time) as any other Extra-Terrestrial threat. We are not prepared at all for it and it is a dumb threat. A threat from a smart ET would prove even more difficult to overcome, especially with no means of traveling to the threat before it is on top of us. If an alien attacker only wished to destroy us, they could manipulate the orbit of a larger NEO to impact Earth and wipe us out. There is little humanity could do about it. And what if the aliens decided to use a NEO to wipe us out? There is still little we could to about it.

With respect to technology development we can not afford to take the approach of the victims of Indian attacks in western novels and movies. We can not expect the technology cavalry to come riding over the hill just in the nick of time as we run out of ammunition. We would be much better served to begin preparing now. PREPARE NOW, SURVIVE LATER. We can make advanced technology weapons that are force multipliers in conventional combat scenarios so that we do not have to support the costs for two separate sets of weapons.

2.5 Unconventional Considerations

Linguistics

Popular folklore would have us believe that any alien advanced enough to come to our planet would also be advanced enough to immediately be able to interpret and emulate all the languages spoken on Earth. In reality it is unlikely that any ET would have communication technology that is compatible with ours. ET could probably not phone home with our technology. Nor could he phone us with his. He would have to have incredibly large radio antennas or have extremely high power electronics to communicate with us or with home from Earth. Optical communication systems as we know would offer little if any advantage. Therefore we should anticipate that their communications technology would also be significantly advanced over ours.

As we contemplate exploration of the universe, there are a variety of technologies that we need to be considering. Technology to communicate back to Earth and Earth to the Explorers is among those that need to be developed. Due to the limitations of light speed communications, breakthroughs in communications technologies should be sought. Technical development is quite possible without accompanying linguistic development. We have seen little change in our linguistic skills and abilities over the last couple of centuries. Perhaps the last significant linguistic advancement was the written alphabet. There are many creatures on Earth that have minimal levels of specie specific communications such as barks, meows, chirps, etc. Yet in 3-5 million years we have not learned to speak with them. Although we have trained some animals in human languages, we have yet to decipher theirs. This could be a good starting point for learning the patterns of nonhuman language which would be helpful for communicating with ET – before and after visitation. However, we have seen tremendous technical development in that time frame. The two appear to have very little to no correlation.

This seems very likely to be the case for an ET who was technically developed enough to arrive at earth. It would appear that the ability to communicate via an existing Earth language would perhaps be an accident at best. Of course it is possible that an ET would have the foresight to be developing linguistics along with technology. Linguistics is certainly an area that we would be wise to develop in preparation for any galactic exploration. Even if we only prepare for a visit from ET, we should be expending significant effort in the area of linguistics so that we could be the interpreter and emulator as opposed to ET having that burden. The ability to communicate with an ET who had not decided whether he would be hostile or friendly might be the deciding factor in achieving a peaceful outcome. Also, if we cannot decipher the ET language, are we to blindly trust an ET interpreter?

The then Assistant Director of Central Intelligence for Analysis and Production said on September 15, 2003 "along with everything else, our language requirements have changed. We need people who can speak Dari, Pushtu and other languages once considered exotic but now deemed essential. Here we sometimes run into difficult managerial decisions: how do you free up analysts to learn languages when you also need them to continue working as analysts? It takes an analyst 33 months to become fluent in a non-Roman alphabet language. That is a long time to be "out of the line."

ET will probably speak a non-Roman alphabet language. We will be in for a long learning period should ET arrive. Can we survive 33 months or longer while we learn to communicate with him?

An Alien Linguistics Exercise

Suppose you want to talk to aliens. Assume only that the aliens are as advanced as we are, not magically or historically better or worse. Below is a possible approach that a mathematician (universe wide most likely) would start in teaching a new language. ET would start with the primer for the language. It might look like this: (Note that we are taking the liberty of using symbols on a normal keyboard to represent alien symbols).

!$!?!
!#!?!
!@!?!
!%!?;
!$~?~
~$!?~
~@~?~
!@~?!
~@!?!
~#!?~
!#~?#~
!%~?!
~%!?;

A mathematician will look at this and say, "Hey, this is a primer on a language!" Here is why, starting from the top. !$!?! Here we have a simple mathematical expression, would be a mathematicians hypothesis. Assume that aliens would start counting with zero and try that. Therefore, 0$0?0 Hey! if we look at all the sequences the ? comes in the same place! Could it be an "equals" sign? 0$0=0 could be, but the $ could be any mathematical operation since 0 plus, minus, or multiplied by itself is zero.

Hmmm, let's press on.

!#!?! 0#0=0
!@!?! 0@0=0
!%!?; 0%0=;

Hey, there are just enough symbols here for plus, minus, multiply, and divide. But which is which? If you are clever enough you will catch that the last of the four tells us something, but we will come back to that. We might be on the right track if this alien speaks math.

!$~?~ 0$~=~ is not enough to go further and more data is needed. So consider, ~$!?~ ~$!=~. Aha! ~ is some number maybe 1 but we can't be sure, but it is a number. And $ is the "plus" sign. You see why? Any number plus zero is that number and zero plus any number is that number. That is what these two equations say!

0+~=~ and ~+0=~ (oh the little ~ is called a tilde, or as a Chinese physics professor of Dr. Taylor's used to say "twiddle"). Ok, what is ~ in the alien's language? Look at the next equation ~@~?~ ~@~=~ or some number some math operation the same number equals the same number. Hmmmmmmmm, ah! 1x1=1 How about that? Maybe but we need more. Let's assume that for now this is true. Then ~ means 1 and @ means multiply if we are right.

The next equation, if we are correct, will read

!@~?! 0x1=0

This is true!

The next equation, if we are correct, will read

~@!?! 1x0=0

Again true! Good!

The next equation, if we are right, will read

~#!?~ 1#0=1

Okay could be plus or minus at least, maybe other things. Need more. Add the next equation !

#~?#~ 0#1=#1

Eureka! 1 minus 0 is 1 but 0 minus 1 is minus 1 or
1-0=1
0-1=-1

That must be it!

Last two equations:

!%~?! 0%1=0

Could be a lot of things again but since we have learned plus, minus, and multiply we can see where it is going. But to be sure we need the last equation ~%!?; 1%0=; Hmmm, let's assume % could be divide. Therefore the last two equations become 0/1=0. Yes this is true, and 1/0=; Ha! Then the semicolon must be infinity! We know that any number divided by zero is infinity. This would let all of the equations work properly including the fourth one that I said I would come back to! The fourth equation now also would read

!#!?; 0/0=infinity

and

~#!?; 1/0=infinity .

Let's summarize now. You say this didn't teach us an alien language. Correct, it did not teach the whole thing, but it is the start of it. A complete language would continue in the same manner as we have been going. But here is what we did learn: the alien symbols for zero, plus, minus, equals, multiply, divide, one, and infinity.

In just a few minutes we have learned 8 alien words simply from 6th grade level math. There are actually several other words in this example. Such an approach may work for mathematical terms; however, it may not be well suited for learning terms for emotions such as angry or melancholy. Terms for concepts such as hunger, competitive or great may take an entirely different approach altogether. It might be useful to use movies, sounds, pictures, and objects as examples for other concepts. An example would be the word "touchdown." When the National Football League (NFL) created the NFLEurope it was quickly discovered that several of the European languages had no word for the concept of a touchdown in American football. In this case the Europeans use the American word "touchdown" and from example know what its meaning is.

We should also note here that Carl Sagan makes similar use of this type of linguistics and communications process in his book *Contact* - although, we go into more detail here.

First Contact Plan – What Do We Do First?

Any plan of action is dependent on the human race's ability and willingness to communicate with the invader. However, since we cannot communicate with any other species on earth why should we think we could communicate with an alien? Figure 2.19 shows a possible configuration for a decision flow process. Of course the process is purely notional. The only public protocols are those set aside by the SETI organizations known as the SETI Protocol. But the SETI protocol is simply a handshake agreement between SETI astronomers on how they would handle knowledge of an alien civilization. There is no legally binding agreement to it.

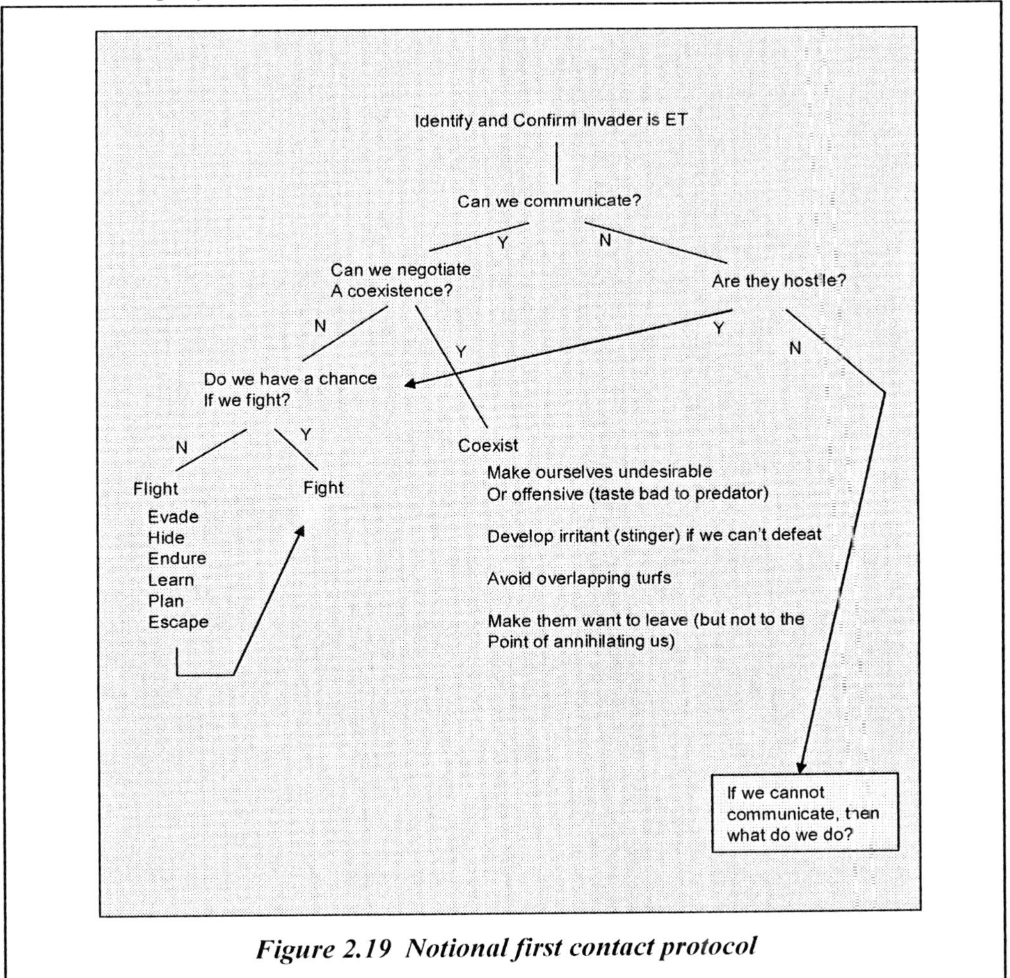

Figure 2.19 Notional first contact protocol

There is a real quarantine law also known as the "alien exposure law". This law was put in place before the Apollo missions and was intended to prevent the accidental exposure of the general populace to an alien biohazard. But there is no real protocol for the first actions following alien contact/invasion – or at least not that we are aware of. More serious development of such protocols should be considered. Figure 2.19 is just a starting point for this and is only intended as such.

Theological Warfare & Effects

How will the general population react to an invasion? Will they even believe it is happening? How will their beliefs and responses aid or hinder the overall defense of our cultures and the planet? How will religious, moral and ethical beliefs influence the decision making process? Should governments disclose evidence of invasion if it is not obvious or withhold it in an attempt to control the masses? These types of issues are the topic of this section.

To begin this discussion we point out that this is a thought experiment as an effort to increase our preparedness in the actual event of an alien invasion. It is not an effort to document, substantiate or explain events that are claimed to have already occurred. If you are interested in such accounts I would refer you to Michael Craft's *Alien Impact* or Patrick Huyghe's *Field Guide to Extraterrestrials*. Again, this book is meant to be a serious textbook on the development of strategies if an ET invasion ever occurs. This discussion is purely a planning exercise much like the Cold War threat and battle scenarios for Soviet invasion of the US.

There are many examples from history that provide insight into the responses to exploration, invasion, attack, or, worst of all, harvesting. One example is the early explorers to the Americas and the profound effect that their technology, religion, diseases and new animal species had on the natives. The destruction of an entire civilization due the Spanish quest for gold is an example of invasion. The U.S. response to the infamous September 11, 2001 terrorist attacks on New York and Washington D.C. are examples of unexpected, un-prepared for attacks. The capture, transport, sale and enslavement of Africans for the US and other slave markets are examples of harvesting. While these are all examples of a terrestrial foe, the response of the governments and the effects and survival of the populations provide good case studies for preparing for an enemy of any origin.

Why are religions or morals an issue at all in this discussion? Take the examples above. How did the perpetrators justify their actions and not consider their activities immoral or unjust? During the conquest of North and Central America, the natives were considered to be un-civilized, heathens. They were considered to be soulless animals and therefore it was acceptable to do anything you wanted with them or to them. The same was true for the slaves transported in massive "alien" ships from Africa to the shores of the US. Why did millions of protestors across the globe rise up against the United States for going to war after 9/11 even though they acknowledged that the US had clearly been attacked? The answer to all of these questions is, "because of their religious or moral beliefs". Some may debate this last point and suggest that the U.S. actions in Afghanistan were supported worldwide but other Global War on Terrorism actions such as in Iraq were not. The point is still valid in either case. Our actions are dictated by what we perceive as right and wrong, good or bad.

How do we decide what is right or wrong? Our governments and religions make laws and rules that we are obliged to abide by in order to have a relatively stable society. We are trained in them our entire lives. When the actions and decisions of government or religious leaders are in conflict with the beliefs of the masses, the society becomes unstable. This is why it is important to consider these issues when forming a defensive strategy.

We often burden ourselves with deciding whether it is right or wrong to take action against an invader before we do anything else. How did the American Indian's system of beliefs influence their decisions to fight, flee or be assimilated? Did controversy within their tribal units slow their response? Did competing tribes recognize the white man as a technologically advanced species, determined to take their "planet," and quickly band together in unison to oppose being conquered? Did the African nations and tribes band together and stop the invasion of their continent and harvesting of their populations? No, in some cases, they even raided competing tribes and sold prisoners to the "alien invaders." Again, religious, moral and ethical beliefs allowed individuals or sub-groups to justify their actions even though it should have been clear that their actions were detrimental to the whole. Disagreements over religious beliefs, gods and the advancement or preservation of individual cultures prevented unions that could have succeeded in the defense of whole continents.

What can we do to prepare ourselves for invasion? How can we benefit from the beliefs of world religions to unite and preserve our existence rather than impede it? Where will our beliefs cause problems or delay critical actions? What and how can we make changes, now, that will enable us to be decisive and effective if the need arises?

Analysis

If Earth was invaded by an alien species there would be a variety of reactions from the general population depending on the situation. The reactions and effect on our ability to defend ourselves will vary greatly depending upon many factors. Table 2.1 lists some of the possible factors. These can be analyzed by taking one from each column to create different scenarios and determining the best defensive approach for that set of circumstances. The result would be the information to create a decision tree for the decision makers or for individuals in their "survival manual."

Table 2.1. Comparison of Government and General Population Probable Responses to ET Invaders

Invader	Government	Population
Hostile Predator Type	Prepared for Invasion	Not aware of threat
Benign Explorer	Unprepared	Apathetic
Godlike	Openly Hostile	Sympathetic to invader
	Secretly Hostile	Hysterical
	Sides with invader	Hostile to invader
		Worships invader

Ideas & Concepts to be analyzed and expanded on by the reader or student:

- Many wars are justified by religions if not caused by them. If aliens invade it may not be because they are hungry or need our resources but because they are missionaries and must spread their religion to the heathens of the galaxy. We are the smartest things we know of in the universe and some of our fine brethren flew airplanes into buildings on September 11, 2001 with hopes of being rewarded in heaven for killing their own species. How do the mass deaths resulting from the Crusades relate? How do we prepare ourselves for being converted? Do we refuse? Do we profess our own religious beliefs?
- The soul by religious definitions could be misconstrued from what might actually be the ultimate spacecraft (this is not an original idea, Q from *Star Trek:TNG* is a good example). Pure energy level aliens at α_5 levels or higher might appear this way. If such travelers were confirmed how should they be presented to leaders or the population? As deities or aliens? Which would be easier to sell to a group, population or leader?
- Good vs evil is the foundation of many religions with parallels in the physical world - light/dark, matter/antimatter, etc. Logically we can expect good aliens and bad aliens. Will it be ok to kill bad aliens but not good aliens? We must be prepared to make this discrimination. Or are good and evil only human concepts? Do cats think that dogs are evil? Do mice think that cats are evil? Does cheese think mice are evil? We might be required to commit genocide on an entire alien race simply to ensure our own survival. Is that an evil concept or is that merely a survival concept?
- We already have religious cults waiting for aliens to take them away. What will happen when aliens really do show up? How many folks will "walk on the ship" and join them? Will these cause problems within our defensive tactics and strategies?

- There are those who still do not believe we went to the moon. It is likely that these individuals will not believe in aliens either simply because they see an alien on TV (This is a self correcting problem)?
- Stages of cultural assimilation from a sociological standpoint are: reaction, rejection, experimentation, acceptance, cultural change. These have been studied many times throughout our history. How can we learn from this pattern?
- To survive we must continue to meet Maslow's hierarchy of needs. Body, security, social, ego, self-actualization. What is the minimum we need to do to achieve this?
- Very advanced (magical) or godlike invaders will disrupt our religions and systems of moral beliefs. Or, perhaps, they will become part of them. This needs to be introduced and considered, now, in our theologies as a part of any defensive strategy. Political and religious leaders need to understand the issues and be prepared to lead the masses without hysteria. We do not want to end up with a feudal religious based and controlled civilization such as the Goa'uld of *Stargate SG-1* or the Vorda and the Changelings of *ST:DSN*. How do you get mainstream religions to consider the possibility of aliens posing as Gods in their theology?

When faced with overwhelming defeat and annihilation, subservience to the point of worship may be a good survival strategy (at least for the short term, until we can understand enough to defeat or decide to accept the invader). Heinlein's *Sixth Column* offers a similar concept. Could you accept that? How would you convince others to do so?

Deception & Denial

A tactic that may be quite useful in surviving an unfriendly alien counter is deception and denial, commonly known as D& D. Deception and denial are closely related with the distinction between the two sometimes being difficult to recognize. According to Webster's Dictionary deception is "the act of deceiving; illusion; fraud." Deception is fairly easy to understand; it is simply trying to fool another being. It typically involves the use of visual effects. It involves the use of such tactics as hiding tanks inside a cardboard structure. The cardboard structure is there to prevent an outsider from knowing the whereabouts of the tank. The use of decoys is another formal deception.

We've seen old Disney Davy Crockett movies where Davy and a few cohorts would place rifles around a group to try to convince them that they were surrounded by a superior force and hopefully get them to surrender. This was a tactic also credit to Sergeant York in World War I. And there have been several instances in history where military forces have used inflatable aircraft, tank, and other armored vehicle decoys for deception. Most modern inter-continental ballistic missile (ICBM) payloads employ warhead decoys along with the actual warheads to confuse radar signatures and antimissile systems.

Denial according to Webster is "a refusal of a request; refusal or reluctance to admit the truth of something." And denial implies false response to interrogation or inquisitive probes. An example of deception might be to hide one communications signal within a stream of communication signals. One might place a low-power transmitter for telephone communication beside a high-power television transmitter using a frequency very close to that of the television transmitter. One might use such RF techniques has spread spectrum and frequency hopping to deny the collection of one's signal. The use of encryption has been a well-known tactic for denying one's voice communications.

2.6 Chapter Summary

This chapter was designed to offer the reader a brief understanding of modern warfare and how we might possibly implement our ability to make war against a significantly advanced alien force. Section 2.1 discusses the twelve basic principles of warfare and how they might need to be redefined due to an alien invasion. Also in this section we offer some basic combat modeling and simulation to illustrate the impact of force multipliers and force replenishment. The outcome of the simulations suggests that a reserve force would prove to be a crucial element in the survival of humanity.

The reserve force would most likely have to come from any surviving civilian population. Section 2.2 discusses the need for better understanding of civil defense in order to protect the civilian population and therefore the reserve component. If a civil defense system is successful then the surviving population will most likely be needed in the asymmetric war that would follow the initial invasion. Aspects of the asymmetric war are covered in Section 2.3.

In Section 2.3 a discussion of the asymmetry between an advanced alien force and Earth forces leads to the analogous situations in the GWOT. Other lessons from history such as the Soviet-Afghanistan war should be considered.

Section 2.4 gives some examples of technologies that exist now that might have some significant impact to the alien forces. There is also a discussion of some tactics in combating shields. Other science fiction technologies such as Powered Armor Suits and Mecha are also discussed. An initial discussion on first contact reactions was given in order to spark future discussion.

Section 2.5 discusses unconventional considerations that must be addressed. The ones discussed were linguistics, theological warfare, and deception and denial. This section is not an exhaustive discussion of the unconventional considerations, but merely a starting point.

3

Motive Based Classifications of Extra-Terrestrials

We discussed in Chapter 1 the power classification of the ET civilizations based on their warp factor ability. In this chapter, we will discuss the ETs in a different way. Here we will consider the motivation of the ETs and why they are attacking Earth. When it comes to understanding the alien philosophy or motivation, it would seem that very little science can be applied and therefore, discussion of the sorts of enemies that Man may face in the future must for the present be entirely speculative. We know nothing at all about intelligent alien life. We are not even sure that it exists, although its nonexistence may seem very improbable.

But speculation must be constrained by what we know of physics (a great deal, and we can be sure that our knowledge is as valid in the Andromeda Galaxy as it is on Earth), astronomy (still a great deal), and biology (quite a bit, although all this knowledge is confined to the special case of Earth biology, and we do not know which features of Earth biology are universal and which are not). As for psychology and sociology...some might ask how much real knowledge we have of our own psychology and sociology. It appears to be quite limited.

But what we will do is to use science fiction examples where appropriate to discuss types of alien motivations. It is often best to explain a concept through example and there are a plethora of examples of possible aliens in the science fiction community. Therefore, becoming a student of science fiction stereotypes and possibilities seems to be a potential tool in classifying alien motivations.

3.1 Motivators

Before we deal with intentions it is important to look at motivators. The human race knows only two motivators. Those are 1) desire for gain and 2) fear of loss. Regardless of what we say motivates us; that motivating factor can be placed into one of those two categories. Unlike technology where it is probable that ET is more advanced than we are, it is difficult to believe that there are more than two monitors in the universe. Therefore it is logical to believe that ET is driven by the same set of factors that drives the human race. This discussion will proceed based on that premise. Obviously, we do not know what may be ET's real motivators. We can only postulate ET's motivation. We will make an effort to to identify some motivators in each category that could drive ET to visit the earth.

We will first examine possible motivators that represent desire for gain. He could be driven by one or more of the following desires for gain or by a desire for gain that would be something completely foreign to us. He could be driven by something we would not recognize as a gain.

- He could come trying to gain resources. Those resources might be resources that are becoming scarce on his native planet or resources that he has detected through some sensing technique. The resource may be new to his planet or a rare commodity in his world but ET has decided for some reason that it offers a great advantage of some sort to his planet.
- He could come to gain a labor source and a resulting economic advantage. He may be looking for slave labor or looking for cheap labor or he may simply be looking to augment the labor force on his planet. He could see the human race as being available and educable. In addition to being available he may see us as being quite vulnerable and easily captured. Therefore, humans would be satisfactory for performing menial labor tasks to support his race.
- He could come looking for food. Humans could be that food source. He is unlikely to looking for our agricultural products as they are unlikely to survive the trip back to his home planet unless he has developed storage and transportation technologies well beyond our grasp. It is certainly a reasonable probability that he will have learned such techniques as a requirement for undertaking such a harvesting trip. We will need to develop such technologies if we become the intergalactic traveler as a means of feeding our population.
- He may come to gain a place of refuge from authorities or competing factions on his home planet. He may be a fugitive from Justice seeking asylum. If that were the case he would possibly have to rely upon random selection of earth as his new home and hope that his pursuers did not make the same random selection in an effort to find him. It is also possible that he would merely be an outcast or group of outcast that would settle on earth as a new home. *(The Hidden* and *I Come in Peace!)*
- He may be driven by desire to gain ethnic cleansing. He could be much like Hitler and his belief in the Aryan race. He could be out to eliminate all ethnic groups other than his own in the universe.
- He could be like the Pilgrims who came to the US. He could be seeking religious freedom.
- He could be looking for new sexual partners. This desire could be experimental or it could be to avoid the weakness of the populace caused by inbreeding. It is

possible that he is just bored with the current sexual options available on his home planet.

- He could come seeking his fortune in the Earth's resources.
- He could come seeking economic advantage in the form of land to settle on and earn a living like many of those who colonized the American West.
- His motivation could be as simple as basic as greed. That is certainly a motivation that we humans know quite well.

It is also possible that his motive could be a fear of loss. While there is a shorter list of things that are fear of loss, we will attempt to list here possible motivators in this category.

- He could fear the loss of his environment. That might be driven by reduction in the beneficial gaseous content of the atmosphere. Water could be becoming scarce. The nutrients into soil could be depleted to the point that agriculture is difficult. These are environmental factors that we humans may certainly face in the foreseeable future.
- He may fear overpopulation of his planet. Overpopulation could make it to difficult to provide sufficient food on the land available to ET. He would run the risk of losing the possibility for the expansion of his society. This could force him to look for new lands to colonize just as the Europeans colonized the Americas.
- He might fear the loss of his civilization without allies in an ongoing battle (like the Galactic in Ringo's *Legacy of the Aldenata books*)

We believe that it is possible to see fear of loss drive the desire for gain. As in the case of the fear of loss scenarios described, in order to have the society survive and prosper the populace would be driven to have a desire to gain stability and prosperity. In each of those cases ET would be driven to gain new resources and expand his civilization.

We think that it is unlikely to see desire for gain drive fear of loss. It is possible but highly unlikely. It appears that desire for gain motives significantly out number fear of loss motives. As a result we should expect there to be a higher probability of seeing an ET that is here to satisfy a desire than one would that is afraid of losing something. The actions of ETs driven by desire for gain are likely to be hostile or detrimental in some way. Following this assumption logic, the probability of being visited a hostile ET appears greater than the probability of a visit by a benevolent ET. This is a good reason to have SETI stop broadcasting to any and every ET regardless of motives or intentions. We believe that it would be good for SETI to be attempting to find ETs via passive techniques. Another reason it would be good to stop SETI broadcasts is that it is not a sanctioned organization to deal with ET.

About all we can say for sure about potential invaders of Earth is that they must have space flight technology or at least have the services of a race that does. Microorganisms that reach Earth in meteorites or micrometorites can hardly be considered invaders, although there is a remote possibility that they could wreak havoc by causing disease. (If we can judge by Earth microorganisms, alien viruses could not even reproduce in the bodies of Earth species; alien bacteria and protozoa might survive in Earth species, but it is most unlikely that they would be pathogenic.)

What is required here is imagination. All of the authors of this textbook agreed during various discussions that the science fiction community has already developed a very large amount of work on the possible ET motivational scenarios. These range from the most primal (ET is hungry) to the most bizarre (the Earth is in the way of an intergalactic highway).

In this chapter, we will discuss the scientific likelihood of various classes of ET motivation and we will refer to the science fiction arena to describe them. As an example, consider the giant sentient robot Unicron from the *Transformers ™* cartoon. Unicron is a giant planet-sized robot that eats planets for energy from which he survives. When Unicron gets hungry, he eats a planet. We would classify Unicron as a Destructor type ET. His nature is to simply destroy. Therefore, Unicron would be classed as a Destructor. It is possible to argue that Unicron is a Strip Miner since he is destroying the planet for its resources. But the outcome of his efforts were always to purely devour a planet. Strip Mining would imply that there might be something left when the ET was gone. There are gray areas in this type of classification and further study could be made to clarify all the rules and requirements for classifying an alien as one thing or another. That study is beyond the scope of this text, but offers an interesting discussion topic for the readers.

3.2 Destructors

Fictional examples: *Titan AE*, *Mars Attacks*, *Earth vs. the Flying Saucers*, *Unicron* from *The Transformers*, Galacticus from *The Fantastic Four*, Berserkers, and the Replicators from *Stargate SG-1*.

Other science fiction examples of Destructors are the seemingly insane Martians from the movie *Mars Attacks*, the planet eater Galactus from *The Fantastic Four*, the Drej from the movie *Titan A.E.* (again there might be a discussion here as to the Drej motivation), and there are many other examples.

One can imagine invaders who have no apparent motive other than destruction. There are very few examples of such invaders in human history. Human armies may be ruthless in the extreme, and they may derive psychological benefits from wanton slaughter, but they almost always have objective purposes as well...usually looting or acquisition of territory.

There are very few animal species whose members kill except for food or in defense of themselves or their social groups. (There are species some of whose members kill wantonly under special circumstances, as in the case of the bull elephant during mating season.) Bloodthirstiness is usually not a species survival trait, and is therefore selected against.

Natural selection is much less effective in species having advanced technology than in those on a purely animal level. In that case natural selection will be dominated by mental acumen – the ability to make the best use of the technology. But before a species can acquire advanced technology, it must survive its primitive beginnings, and a truly murderous species is less likely to do so.

Nevertheless, there are possibly races that are prepared to cross interstellar space, or at least send their machines across interstellar space, merely to destroy. They are probably rare, but they may account for a significant part of the warfare in the Galaxy.

There may also exist military robots that have, for one reason or another, exceeded their programming, and are attacking any intelligent beings and any other robots that they encounter. (Cylons from *Battlestar Galactica*, Replicators from *Stargate SG-1*)

One cannot negotiate with beings whose destructiveness has no rational basis. Such beings might or might not withdraw short of the complete destruction of their forces.

3.3 Assimilators

A different type of alien motivation might be that of an Assimilator. Assimilators have the motivation of making all other living creatures part of their society or culture. The Borg from *Star Trek: the Next Generation* is a good example of the Assimilator type alien. Another great example are the Slugs from Heinlein's *The Puppet Masters*.

The point of using the popular science fiction images is to explain the category without having to go into great detail. We cite several popular examples of each classification type with the hopes that the reader has been exposed to at least one of them. If the reader is interested in studying more detail of the endless types and scenarios of would be attacking aliens they can refer to the movies, books, comic books, and television series that we reference here as an example of the usefulness of Pop Science Fiction (Rocky Horror argument). Otherwise, Table 3.1 shows a list of the types of alien motivations along with some popular science fiction examples.

From Table 3.1 we now have a means of classifying both the power level of the ET adversaries as well as a possible motivation or alien type. This table is by no means to be considered exhaustive. It is likely that many possibilities that we did not think of exist. However, Table 3.1 is a good starting point. Also, the categories may be general enough as to be mostly encompassing of the possibilities. Infinite subsets of the existing motive classifications probably exist. Notice from Table 3.1 that the majority of classes are Hostiles ET's. This might indicates that we have the greatest probability of being visited a Hostile ET. On the other hand, it might just be the types of aliens that we could imagine are limited. Are there many other types?

Table 3.1. Alien Motivation Classifications

Alien Type	Power Level	Popular Example
Destructors	α_1 - α_3	*Titan AE, Mars Attacks, Earth vs. The Flying Saucers,* Unicron from *Transformers,* Galactus from *The Fantastic Four*
Conquerors	α_1 - α_3	*Battlefield Earth, War of the Worlds,* Klingons, Spectra from *Battle of the Planets,* Ra from *Stargate*
Strip Miners	α_1 - α_3	*V, Independence Day*
Harvesters (including Business Men)	α_1 - α_3	*V, Signs, the drug dealer from* I Come in Peace
Hunters, Collectors	α_1 - α_3	*Predator,* The Preserver from *Superman*
Assimilators	α_1 - α_4	Borg, Purity from *The X-Files, The Puppet Masters, Invasion of the Body Snatchers, The Faculty, Dark Skies*
Opposing Warriors, Location and happenstance	α_2 - α_4	*Robotech, Hitchhikers Guide to the Galaxy, I Come in Peace, The Hidden, Superman and Doomsday, Superman and Darkseid,* Autobot/Decepticon war on *Transformers*
Tourists	α_2 - α_5	*Men in Black, The Blackhole Travel Agency*
Meddlers	α_3 - α_5	Q, *The King James Version of The Holy Bible,* All of the ancient mythical Gods, *Waiting for the Galactic Bus,* Gaurdians from *The Green Lantern, The Day the Earth Stood Still, They Live*
Movers	α_0 - α_3	*Independence Day, Orphans of the Sky,* Drej from *Titan A.E.*
Experimenters (Meddle in physiology not just sociology)	α_2 - α_5	*Taken,* Grays from *The X-Files*
Experiments (actual experiments trying to conquer)	α_0 - α_5	*Species, Evolution, The Thing*
Terraformers	α_3 - α_4	*Von Neumann's War,* Replicators from *Stargate SG1*
Motivation Unknown	α_0 - α_5	No clear motivation could be determined,

We now will discuss a little more about some of the alien type classifications given in Table 3.1. Again the Table is not necessarily exhaustive and the details and classification rules could be defined more precisely but it is a good starting place.

3.4 Conquerors

Fictional examples: *War of the Worlds*, *Battlefield Earth*, Klingons, *Battle of the Planets*

Conquest is undoubtedly the principal motivation for the greater part of human warfare. Many animals, solitary and social, are territorial, but wars of conquest seem to be unknown among animals.

Conquerors always want to acquire territory. They may or may not want to acquire the possessions of the conquered.

If they are on a much higher cultural level than the conquered, those possessions are probably of little value to them, as in the case of the European conquerors of Africa and the New World. The conquered people may be of value as a source of labor, but if the conquerors are sufficiently advanced technologically, such labor may be of no use to them.

If the conquerors are on a much lower cultural level than the conquered--which is occasionally the case--they may not consider the possessions of the conquered to be of much value, and even if they do, they may not be able to maintain those possessions; consider the barbarians who overthrew the Western Roman Empire. Relatively primitive conquerors, if they want to maintain the more advanced culture of the conquered, may avail themselves of the services of the conquered in order to do so. If they do, they risk assimilation by the conquered, as in the case of the Mongols in China.

One can imagine conquerors who find the conquered civilization to be so alien that they have no desire to acquire its possessions--may even find them repugnant, and want to destroy them. This was to some extent the case with the Spanish in Latin America.

If, for whatever reason, the conquered people are of no value to the conquerors, they may be exterminated as a possible threat, even if a minor one. This has been the fate of some primitive peoples on Earth, such as the Guanches of the Canary Islands. In other cases, the conquered are not exterminated, but are brutally oppressed. In yet other cases, they are assimilated by the conquerors; this process is far advanced for the American Indians. Assimilation would be difficult, if not impossible, if conquerors and conquered belonged to different species. If assimilation of the conquered is not convenient, extermination or enslavement of the conquered appear to be the logic alternatives.

Alien invaders might come from a planet with a very different environment than Earth. If their technologies were sufficiently advanced--and if they could cross interstellar space, it undoubtedly would be--they might want to transform Earth's environment into one more congenial to them. If they did, Man--and most, if not all, of Earth's flora and fauna--would be exterminated incidentally. This would be case unless those species adapted rapidly to the new environment.

At this point we do not see the need to describe each alien classification in great detail. Table 3.1 gives enough popular examples for the reader to enjoy classifying various conceptual aliens one way or another.

3.5 Understanding the Alien Motivations to Direct Our Actions

As the visitor or as the visited, we will need to know how to engage ET. We will need rules of engagement. Those rules will vary according to the category of ET as discussed in Table 2.1 and the motivations given in Table 3.1. The comparison of the alien invaders to the data in these tables will also drive the first contact protocols given in Figure 2.19. It would be a shame to treat a Benevolent ET as a Hostile ET. We would risk losing the benefits of that benevolence. It could be a tragedy to treat a Hostile ET as we would a Benevolent ET. We could lose our civilization or worse - our lives including those of the entire population of Earth. This is the fallacy of the SETI approach of welcome one and all ETs to Earth, especially without preparation. Which, if any, ET will heed SETI's invitation? We are unable to predict at this time, but we'd better hope for the best. We realize we mention this several times throughout this book, but there seems to be no greater point to make than we don't know what is out there so we had better be careful who we invite to come visit us.

Regardless of the category of ET the most significant challenge is that of determining the ETs intentions and then his military capabilities. If an alien invader shows up it might be easier to determine how best to deal with that invasion if we understand what they are after. This is the reason for this chapter. We should start thinking about what the alien wants and what it would take to get him to leave. If the invasion is from a mindless machine army bent on terraforming for future owners (von Neumann probes) then the only motivation for them is to eat raw materials and build infrastructure (and/or self replicate) then move on to the next target. The only defense for that type of invasion is to destroy every single one of the probes. Understanding the motivation will help pick a tactic. A good science fiction example of this is *Von Neumann's War* by John Ringo and Travis S. Taylor.

Figure 3.1 shows a flowchart for the possible decision making process based on the alien power category, their motivation, and including the initial first contact protocols. The initial process is the detection of the alien invader. This might be subtle or blatantly obvious depending on the nature of the invasion. Then a determination of the alien power category is made. The first contact protocols shown in Figure 2.19 are implemented and interim actions are taken. While the interim actions are made a comparison of the aliens to the Alien Motivation & Reaction Plan Database is made. All intelligence, surveillance, and reconnaissance gathered on the aliens is implemented to make a decision as to what type of alien the invaders are and to determine which reaction plan to implement. Following reaction plan implementation is battle damage assessment and modification of the plan if need be. At this point the process is fed back to the motivation and reaction plan determination cycle as interaction with the aliens might alter their motivations.

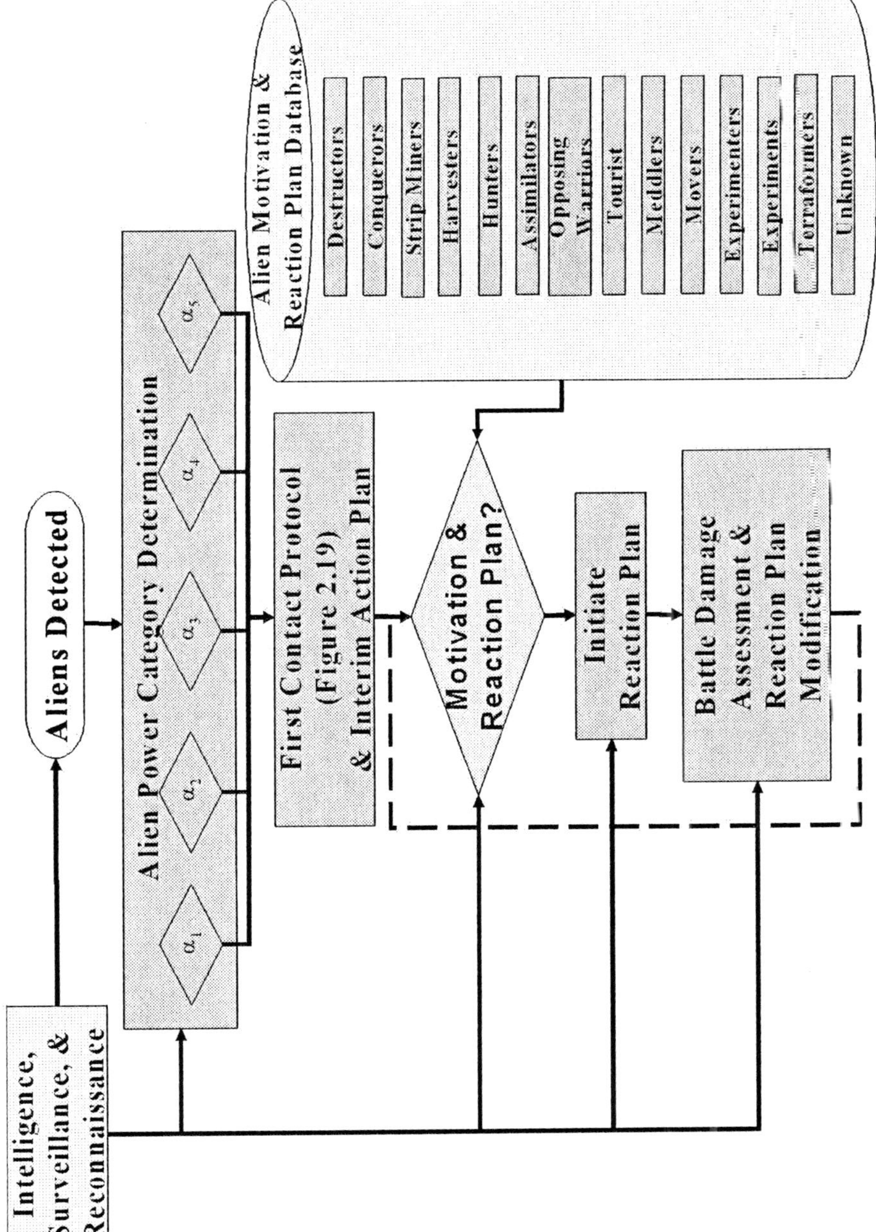

Figure 3.1 Understanding the alien motivation will enable the implementation of a reaction plan!

3.6 Chapter Summary

In this chapter we have discussed various types of alien motivations based on science fiction examples. We discussed in detail in Sections 3.1 – 3.4 specific classification examples. The table that is given within this chapter is to offer a starting point for cataloging these possible alien motivation classifications. It is quite likely that we have left several possibilities out – the list was not meant to necessarily be exhaustive rather than a starting point. Finally, in Section 3.5 we discuss why it is important to understand the alien motivation. Understanding the alien motivation will enable us to choose the proper plan of action if we are prepared.

4

The "Need to Know"

This chapter is difficult to handle emotionally. It requires great responsibility from the readers to understand it. The authors argued and debated this chapter and its implications for months before reaching an agreement on how to address this material. It rekindles the argument of the rights of the individual verses the good of the many as judged by a selected few. You'll see that although on some points the authors disagreed in the minor details of the argument, we agreed on the overall approach and ideology.

The authors would like to make a point before you continue into further sections of this book. We have no information about aliens, UFOs, or any other such event that is not available to the general public. We do not believe in conspiracy theories of any sort and find little logic in most of them.

With that stated we feel it necessary to discuss the issue as to why or why not the public should be told of tactics, defensive strategies, impending invasions or even the mere evidence of alien existence.

As you read this section, please realize that it was written in a blunt and direct manner and was not intended to incite harsh feelings, but rather to stimulate the thought process as to the real issue of "need to know". Granted, we all want to know what happened at Roswell and if alien abductions are real. We hope that this section motivates you and forces you to wonder if you really should "know" and "why". The answer may be yes or no and we think it depends on the individual's specific situation.

It is our strong belief that any knowledge that would give the United States of America an "edge" of some kind over other nations should absolutely be classified at the highest levels. On the other hand, if there is a way to release to the general public the basic knowledge that alien life is a fact without impacting national security, then it is also the responsibility of our government to release that information.

The problem we see is that there are very limited wording methodologies for the release of such information to the world without causing national security threats. If a naked alien body were found in the middle of the desert, with no alien technology with him, should that information be publicized? The fact that such an alien could suddenly appear within our borders might tell our Earthly foes something about our air, sea, or land monitoring and defense capabilities. Again, we must consider if there would be any useful technical information lost or leaked by our releasing such information. If we could perceive that there were no gains, which could be made by our enemies, then we should release the information to the public. Realize here that our enemies might be the aliens not other non-allied nations.

There is also the public argument that "if there is a threat to me I have a right to know". As it stands now, there is no scientifically accepted evidence available to the public that we know about any alien threat, so the aliens cannot be certain that we know about them and are preparing for them. This is also a useful subterfuge. If we released such information to the public then, obviously, the aliens would know that we knew about them. This would enable them to ascertain particulars of our detection offensive and defensive capabilities. It would be better for us if that information were kept secret from invaders and be kept secret from the general public to maximize our ability to protect that information from ET.

Again, we would all like to know, something, anything. But we all do not necessarily need to know for our own protection. We are not at all suggesting that the public could not handle the information. That argument is plausible to some degree, however we have all seen alien attack movies and books for years and are, most all of us anyway, well capable of handling the emotional, philosophical, and theological impacts. But the information gives our nation's military an "edge" on the alien threat just by knowing of it, and releasing that information to the public (likewise to the aliens) removes that edge. For this reason, we suggest that the secrecy is for our own protection.

Preparations for an offensive or defensive encounter with ET are best done by those organizations which are trained for and entrusted with our military capability and defense. They have the weapons, training on those weapons, the manpower, and the technology to provide the greatest probability of success. Besides, there is little or nothing that an individual or uncoordinated group of individuals can accomplish against such an aggressor. One might hide if one knew where ET were going. How long could one hide? If the "streets were empty," this could tip ET off when the element of surprise might be our most effective weapon against ET.

Perhaps there should be a detailed study effort undertaken with the purpose of determining what information is critical to national security and what information if any could be released to the general public. Unfortunately, there may be reason for no information to be released to the public. Perhaps, there may be some. This section discusses this concept further and we hope it helps alleviate some stress of "not knowing". We feel that same stress and are in the same boat with you and that is comforting to us.

4.1 A Gedanken Experiment

Wikipedia defines a Gedanken Experiment as:

In philosophy, physics, and other fields, a ***thought experiment*** *(from the German term* ***Gedankenexperiment****, coined by Ernst Mach) is an attempt to solve a problem using the* *power of human imagination. These experiments are used to attempt to understand something practical through an analogy. Thought experiments design a hypothetical situation in which our intuitive response is contrary to our actual responses in similar real-world situations. Many thought experiments include apparent paradoxes about the known or accepted that with time have led to the reformulation or precision of theories.*

We suggest the following gedanken experiment be conducted to help you understand this problem. Take a poll of a small group of individuals (whether it's from a family discussion, an organizational meeting, a fun party discussion, or a classroom situation is immaterial) as to how they respond to the following scenario.

Scenario #1: An alien spacecraft and a full complement of dead bodies were found just outside one of our nation's Air Force bases. Presently, no details of why the aliens were here can be ascertained. How would your classmates answer the following simple yes or no questions?

Question #1: Should the public be told we have evidence of extra-terrestrial life?

Question #2: Should the public be told we have found alien bodies?

Question #3: Should the public be told we have recovered any alien technology?

Question #4: Does public knowledge create any (even minute) possible threats to national security?

Question #5: Should the incident be kept above Top Secret?

Table 4.1 Contributing Authors Answers to Need to Know Gedanken Experiment Questions for Scenario #1

Individual	Question #1	Question #2	Question #3	Question #4	Question #5
Mr. Anding	No	No	No	Yes	Yes
Dr. Boan	No	No	No	Yes	Yes
Dr. Powell	No	No	No	Yes	Yes
Dr. Taylor	No	No	No	Yes	Yes

We suggest that you compile the data much as shown above in Table 4.1. Then read this section carefully and discuss the questions within the group and make a new table with the new answers in them. Since we developed the question and scenario, we discussed it thoroughly and there is no need for a second table here. This is our post discussion result.

You might try to add other scenarios to the process and see if other scenarios cause different results. Another example scenario is given below.

Scenario #2: An alien mummy is found in an Egyptian Pyramid. The mummy is completely intact and appears to be an Anubis. Other than clothing, the only article of interest in the sarcophagus is a flashlight. The flashlight is nothing fancier than a modern two-dollar battery operated incandescent bulb flashlight and a plastic toothbrush. The mummy and its artifacts are found to be ten thousand years old. Presently, no details of why the alien was here can be ascertained. How would your classmates answer the simple yes or no questions this time? The responses of the authors are shown in Table 4.2.

Table 4.2 Contributing Authors Answers to Need to Know Gedanken Experiment Questions for Scenario #2

Individual	**Question #1**	**Question #2**	**Question #3**	**Question #4**	**Question #5**
Mr. Anding	Yes	Yes	Yes	No	No
Dr. Boan	Yes	No	No	No	Yes
Dr. Powell	Yes	Yes	Yes	No	No
Dr. Taylor	No	No	No	Yes	Yes

Dr. Taylor's answer remains that the alien's motive is unknown and therefore could be a detrimental one for the human race. Until the Anubis' motive for being here is determined, we should not let his fellow aliens (are they watching us?) know that we know they were here. Their plan might have been to manipulate our history in a bad way. We should not reveal that we are close to understanding this. If it turns out that the mummy's motivation for being here was harmless, then the public can be told. Of course this experiment assumes that somehow the U.S. exported this Anubis somehow from Egypt without the knowledge of the Egyptian government and that brings up a whole other interesting debate that we will not get into here.

Mr. Anding believes that in Scenario #2 that the possibility of imminent threat from an ancient dig is low compared to the benefit that could be gained by informing the public and gaining their support for the whole concept and philosophy presented in this book. Physical evidence is necessary to complete the scientific process and move from hypothesis (or speculation) to proven fact. Some facts certainly need to be classified (such as the technology in Scenario #1) but keeping people in ignorance may not provide them protection. Hiding such a monumental discovery as Scenario #2 would withhold the very spark that would ignite an unprecedented wave of enthusiasm, investigation, discovery, support and preparation that would benefit global security. Mr. Anding suggests that the public should be motivated to become knowledgeable in the technical fields required for creating a defense and knowledge base.

Dr. Boan's particular insight here is that the alien body might be spread about in world museums and laboratories and there would be little way of protecting the body and maintaining a stable specimen for analysis. In other words, without the entire body maintained in a closely protected, clean, and research environment there might be "too many chefs in the kitchen, which will spoil the soup" so to speak. One way of preventing this would be to keep the specimen classified where only a handful of known and qualified researchers will have access to it. In some cases, too much help is not helpful.

After further discussion between the authors, Dr. Taylor interjected an interesting point based on Einstein's Special Relativity. A major point that most people will overlook in this topic is Special Relativity. Consider the scenario in *Independence Day* whereas in 1947 an alien spaceship crashed in Roswell and fifty years later the Mothership shows up. To us on Earth it would appear that fifty years had passed and we had given up on the aliens coming back. But what if they were still on their way here? If these creatures were a sublight species then Special Relativity becomes very relevant (pardon the pun). The first small scout ship could have left the Mothership at near the speed of light but just a little faster than the Mothership was traveling. The little ship crashes at Earth but has the time to send a message back to its Mothership on how would be best to attack us. Fifty years pass on Earth, but for the Mothership, which was slowing down from near lightspeed only a few minutes to weeks have passed due to the time dilation effects of Special Relativity. Now the Mothership has been prepared better and its army of ETs is still rearing and ready. We on the other hand spent fifty years forgetting that they were coming.

Now consider the mummy in Scenario #2. Ten thousand years could still only be a few minutes to months for a sublight but near lightspeed faring species. That mummy could have been sent here to, in some unforeseen alien way, confuse our history to make us more susceptible to conquest. Perhaps by posing as a god and predicting that another god will appear in the future. Or perhaps simply by convincing us that we are alone in the universe and not to ever worry about alien invasion. And when the alien Mothership arrives just a few Relative minutes later, we have been defeated by our very history! With this argument in mind Dr. Taylor will most likely never give different answers to the questions of this exercise, no matter what the scenario might be. The other authors had similar epiphanies.

Scenario #3: It is discovered that the Mayan calendar correctly predicts an alien return to Earth in the year 2012 and it is confirmed their arrival is imminent. You will find the answers given by the authors in Table 4.3.

Table 4.3 Contributing Authors Answers to Need to Know Gedanken Experiment Questions for Scenario #3

Individual	Question #1	Question #2	Question #3	Question #4	Question #5
Mr. Anding	Yes	Yes	Yes	Yes	No
Dr. Boan	No	No	No	Yes	Yes
Dr. Powell	Yes	Yes	Yes	Yes	No
Dr. Taylor	No	No	No	Yes	Yes

Dr. Taylor and Dr. Boan both feel that leaking the information to the public is no different than telling the aliens that we know they are coming. This removes any possible aspect of surprise that might enable a military tactical advantage. If we've known they were coming for years and have been preparing for them and they do not know this, we would in essence be implementing an ambush on them. Letting the public know removes that capability.

Now that you have finished the exercise with the scenarios above let us present our argument. First, there is no public need to know when it comes to alien invaders and the mere survival of the human species. Again, we are interested here solely in the SURVIVAL of the human race, not the intellectual evolution or expansion of knowledge just for knowledge's sake. Those things are noble endeavors but not at the cost of the species.

We would like to note here that in the event of an imminent threat like the Mayan calendar scenario, it is likely that eventually the public must be told and their help enlisted. In such a case the entire world might be put to work on the war effort. This is similar to the efforts described in John Ringo's *Legacy of the Aldenata* books. The initial planning should remain classified simply to mitigate logistics risks and the problem of too many chefs in the kitchen spoiling the soup. And for the reason mentioned above that we should maintain our ambush option as long as humanly possible.

It should also remain classified until the absolute last minute necessary because the team working the problem might develop a solution without needing to give public access. This in turn would enable us to have and maintain technical advantage over our terrestrial adversaries.

Another question is what happens if alien technology is required by other governments? If we have been secretive with ours would the other governments do the same? Most likely so. Even if there were some "agreement" to share the technologies it would be difficult to prove full disclosure were taking place. This issue should be considered further.

4.2 What Would YOU Do? What Could YOU Do?

If you had any of the ET knowledge of the above scenarios or any other scenario, what would you do about it? If the military and intelligence organizations have no clue of defense, their weapons are not powerful enough, and specially trained soldiers and operatives can do nothing, and the most brilliant and trusted military scientists have no clue of what to do, what do you think you will do?

This may sound a little rude and, at first, elitist, but it is the honest and blunt truth. There is very little you can do. There are no shelters that would really help and very little forms of preparation on a small scale that could be done. It is possible that a well-designed bomb shelter may offer some protection. That is debatable however.

So, the argument that you might could aid in the defense of mankind is minimal at best, unless it comes down to an infantry war. By that point, you would know all you needed to. What other arguments could be offered to support a public need to know? You should be told if there are aliens or proof of aliens. No good. If you want proof that there should be aliens out there, go learn a lot of math and physics and astronomy and try to gather it yourself. There is absolutely no doubt that there are aliens out there somewhere. If they are advanced enough to travel to Earth is another question. Mathematically speaking, there must be aliens. You can at least feel good in understanding that. What could you do with the knowledge that your math had been experimentally verified and that aliens had been found? Most certainly the gratification would be emotionally satisfying, but what practical use is that knowledge? This topic needs further thought and debate.

Should you be told if they are coming to Earth with hostile intent? Only when it will mean a better chance for human survival should you be told. If they are really showing up, mutilating cattle, abducting individuals, and putting bismuth on their pillows then what could knowing about it do for you. If you are one of the abductees then you already know. If you are not, then why worry about it? We have just shown you with conservative statistics that there is a distinct possibility that aliens exists. You can decide yourself whether to prepare for it or not. You know that tornados exist don't you? Do you have a tornado shelter? What about Hurricanes? Earthquakes? NEOs? Nuclear War?

4.3 Secret Weapons MUST be Secret!

"But I deserve to know, why shouldn't I be told," one might say. If an alien spacecraft were apprehended in some manner by an arm of the U.S. government and it was turned over to the proper organization, that organization might have a chance of reverse engineering the technology. The gains that could be reached could be phenomenal. If the public were told of the existence of the spacecraft, then other governments would claim that they had rights to see the technology and this could cause a political and diplomatic nightmare. Who should be in control of the technology? Who should benefit? Would you want Saddam Hussein or Osama bin Laden to have the magic alien death ray at their disposal. What about the Russians, Chinese, Germans, North Koreans, and the French? Would these governments invite us to the party or keep the technology to themselves? Would they release the information to the US? The Russians had a very active, highly classified UFO programs prior to the "fall of the wall". Information that was stolen in the early days of governance chaos has been purchased and released to the public. The recently released *COMETA Report*, which is an analysis of UFO sightings conducted by a group of French individuals connected to the French military and academia, have suggested that the U.S. knows more than they are telling the rest of the world and that it is unfair to withhold such technology and information from the rest of the world. Is it really unfair to withhold military technology from other governments?

Now of course this essay is a bit facetious and was meant this way to prove a point. We the authors have no knowledge of any alien spacecraft, Area 51, or government conspiracies. But if the U.S. has or were to gain such knowledge or technology we should maintain it, study it, improve upon it if possible, and by all means not let the rest of the world have the slightest idea of its existence. Heinlein once stated, "A secret weapon must be just that, a secret." We agree. If we tell the rest of the world, then our secret weapon, by definition, will lose its status as a secret weapon.

As an example of this, before the world knew that we had stealth bombers and fighters other nations had no reason to go looking for them. Now that they know they exist, our enemies are updating their technologies and will soon be able to find ways of detecting them. Hence, the stealth technology may become practically useless. We have spent billions of dollars to develop the next technology. The need to upgrade technology has a positive side in that it forces us to continue to do research, to develop and to think. If we release to the public knowledge of our antigravity, army stopping device, stun gun, and invulnerable shield made from alien unobtanium or impervium, then it will not be long before our enemies and detractors find a countermeasure for it. This is the absolute prime reason as to why the general public has no need to know.

Does the general public need to know every minute detail about how to build a hydrogen bomb? Some individuals in the science and engineering fields might figure it out, but the good guys or the bad guys usually recruit them. So, do you need that knowledge? Good for you if you learn it, soon you will be getting a job offer and a nondisclosure agreement from one of the factions mentioned above.

4.4 What YOU Can Do

The basic point of this section is that the public should use their heads a little better. Look up at the sky at night and quit wondering IF they are out there and start wondering how many of them there are and when they are coming to eat us. If an individual really have something to offer, then that person should try hard to develop his ideas in a real and scientific way that is useful and offer them to the community. Refereed papers, books, and journal articles can be written and published. Then the world will know those ideas. If there is a secret government group and that individual is trust worthy enough and his skills/ideas are useful then he will get his chance. Not everybody gets to be an astronaut or a quarterback or a millionaire or President.

Your part as an American Citizen is to be the best American Citizen you can be. Use your skills to support our way of life in the best way possible. If you are a sandwich maker, then make the best sandwiches that you possibly can. This will help with our quality of life.

As far as the rest of the world is concerned, our allies get similar support. Our enemies on the other hand are just that, enemies, and should be treated as such; no weapons for them. No secret technologies should be given any other government no matter how big their donation to the ruling political party is. This need for a national technological superiority goes for all parties and independent thinkers in the country. Support the country not our enemies.

4.5 Conspiracies are Illegal

Finally, if it is possible for the so-called secret government organization to leak through academia that proof of alien life exists, well then that is partly their responsibility. There is no reason to hold back all evidence if it can be released in a way that the general public believes it to be recently discovered, not recently unclassified. Otherwise the fact that the government was keeping alien knowledge classified would be obvious when the headlines read, "Alien technology recently declassified by government!"

The public has this distrust in secret military and government organizations since they have no real knowledge of how these organizations are structured. That distrust should not be fueled further because it is untrue. The distrust stems mainly from Hollywood getting it wrong and from bad individuals doing criminal things. Fortunately, the criminals usually get caught. Unfortunately, nobody holds Hollywood accountable for anything.

On the other hand, there are reasons for not letting people have access to this information as discussed above. The UFO conspiracy folks should understand security methods, protocols, procedures, and laws before pointing fingers and making conspiracy claims. Trying to sneak onto Area 51 to see what is going on there is a violation of National Security and should be treated as an illegal act and dealt with accordingly. It would benefit National Security if examples were made of these so-called investigative journalists that have tried to sneak into bases and have followed employees from these bases for television ratings. There is no benefit in violating our nation's security. Yes, there should be a conspiracy; one to promote National Security and that follows laws set forth for such in the Constitution. So, that wouldn't be a conspiracy at all, just law abiding citizens upholding the laws and doing their jobs.

A prime example set forth by the so-called conspiracy theorists is that even individuals in high-ranking positions cannot gain access to this information. J. Edgar Hoover, the former Director of the FBI, for example, could not get any information from any organization about UFOs. There are memos in his handwriting complaining of his being stonewalled. Either the information didn't exist or Hoover did not need to know it. History has taught us that Hoover had quirks in his lifestyle that could make him a prime target for blackmail. Therefore, he was in direct violation of requirements for gaining access to secure information. It is conceivable that he simply could not get the proper security clearance. The rules for gaining such clearances are public knowledge and can be found on the Defense Security Services and the Director of Central Intelligence websites. Hoover would not and should not have been given access to critical information of any sort.

A similar thing occurred during President Clinton's time in office. Apparently he had an investigation into both the Roswell incident and who shot John F. Kennedy. He was also "stonewalled" on both incidents. History has now shown us that President Clinton had particular quirks of his personal life that could make him a prime target for blackmail. From the definitions of being able to gain proper access and security clearances as discussed above, Bill Clinton would not have qualified. This is not a politically motivated statement; it is merely a fact.

It is actually quite a disturbing fact. The disturbing part of the fact is that elected officials are typically given access to classified materials without the same background investigations requirements as average citizen would be forced to undergo to get a security clearance or access. Did President Clinton see all there was to see as the elected President with a President's access, or was there a super classified caveat in the law about certain levels of security and passing the qualifications for that access? Who knows, although we the authors do tend to believe that there was nothing for Clinton to see.

The bottom line about the laws of this country should be just that, the bottom line. Any American citizen, elected to public office or not, should be held to the same requirements for any type of national security access. The National Security Act should cover this but as far as we can tell does not. We are not saying that the elected officials are not trustworthy, just that somebody should check to see. Being an elected official should not be grounds for being granted a security clearance. Let politicians who do not meet the requirements for a security clearance work in areas that do not have access to classified information. There are plenty of those types of positions available. The constituencies of elected officials need to know if representatives can not get a security clearance through the routine investigative process – ideally, they would have that information before the election.

It is this type of strangeness in our political system and its interactions with the secure world that fuels part of the public distrust. Misinformation is another reason. Not purposeful misinformation due to government conspiracies or cover-ups, which the author's hold no stock in, but rather something else.

The fuel for the public distrust comes from television based knowledge about the fictional government cover-ups. These television cover-ups usually concern people selling secrets or finding ways to cheat a program or project out of money. Many UFO conspiracy investigators have tried to make an argument that Lockheed Martin, Boeing, Westinghouse, and others got stinking rich from the reverse engineered technologies from Roswell.

If that is the case then somebody would have done it. But there was no cover-up just, capitalism. J.K. Rowlings, Tom Clancy, Michael Crichton, Stephen King, and many others got filthy rich writing books, was that wrong? No, it was capitalism. Making money is right as long as it is done legally. If these so-called Roswell conspiracies did occur they were not conspiracies but simply classified programs where businesses that qualified, competed for contracts, and capitalized on them. There are General Accounting Office rules for the so-called "black projects" and very very ethical folks oversee them. Occasionally there is a bad apple in any industry. The ethical members of the society just must do their duty and pluck out the bad apples as they find them.

As far as the television view of the super secret world, in very few instances has Hollywood read the NISPOM or DCIDs (this is discussed in greater detail in other sections of the book) and properly portrayed the good nature and ethical requirements of the individuals that qualify to be involved with deeply secret National Security type projects. It is this perception that needs changed.

If we exercise our rights as Americans and vote into public office good ethical people, then our fears of government conspiracies should be at least kept at bay. And all of these elected officials must be held to the same standards that you or I would be held to. Again, not everybody always follows the rules and occasionally there are incidents. This is not the norm believe it or not. Interestingly enough, the central limit theorem also applies here and the average is not the criminal. In fact, the criminals are out past one standard deviation on the "wings" of the bell curve.

The responsibilities of the secret organizations are to uphold the law, to support the American way of life, and to promote general welfare and security of the country as best they can. These are the absolute responsibilities. It would be nice if at the same time the secret organizations released information that has no bearing on national security. This is allowed for in the Freedom of Information Act. So, if there are aliens presently attacking or interacting with Earth then it should be completely obvious that it is a direct matter of National Security and the general public has no immediate need to know. Otherwise, the super ethical individuals in these organizations would have let the knowledge flow into the public in some manner.

We are certain that the general audience of this book will read the statements like the one above describing the "super ethical" individuals in the super secret world as hard to believe. We then must refer you to the extreme requirements in the NISPOM and the DCIDs. Some of the requirements include periodic criminal and lifestyle polygraph examinations, extremely detailed background investigations, and personal economic disclosure to name a few. We believe the majority of Americans to be highly ethical. Others would be placed in a position where ethical behavior is expected of them. As the old adage goes, "People will never exceed your expectations!"

We hope this section has made you think more about whether you really do have a "Need to Know". Do you?

4.6 Chapter Summary

This chapter deals with the very difficult emotional debate over the "need to know" about deeply classified programs versus national security. Is there a public need to know about an impending alien invasion?

In Section 4.1 we offer a few gedanken experiments to offer insight on the need to know debate. A compilation of the author's thoughts is given

Section 4.2 asks the questions: What Would YOU Do? What Could YOU Do? It is hoped that this section compels the reader to answer these questions as honestly as possible. The answer you finally decide upon might surprise you. Then again, it might not.

Section 4.3 makes the Heinleinian point that a secret weapon must be just that...a secret. By releasing all knowledge about alien technologies, civilizations, impending invasions we might remove an ambush capability that would have been the capability that saved humanity.

In Section 4.4 we actually discussed what you CAN do to help. That is, be the best you can be at whatever it you do. If your talents are needed then you will be sought out.

Finally, Section 4.5 gives a discussion on the public mistrust of the government and the conspiracy theory. Conspiracies are illegal. Following the rules set forth for national security is not.

5

First Response, Second Response, Third Response – Did we do it right? Did we win?

5.1 How We've Handled ET in the Past

The reactions of the world's governments to reports of unidentified flying objects (UFOs), or "flying saucers", may give us an indication of how those governments would respond to the reconnaissance phase of an invasion from space. They may give us lessons in how not to respond if an invasion really did occur.

Here we will not be concerned, any more than we must, with the question of "What are or were UFOs?" The noted UFOlogist Karl T. Pflock, like many other UFOlogists, believes that the number of good reports has declined dramatically since the 1970s. He believes that the UFOs may have been alien spaceships, but, if so, they have accomplished whatever they came to the Solar System to do, and have departed.) Whatever UFOs actually are, for several years many highly-placed people in the US government believed that they were alien spaceships. How did the US government and other governments react to this perceived possibility?

We will not discuss most of the "classic" cases, because, with some exceptions, they are more relevant to the question of "What are UFOs?" than to the question of how well governments dealt with the UFO problem. (A "classic" case is simply one that is particularly well-known to UFOlogists. Not all of these cases are really impressive. In fact, some of them are generally agreed to have been explained. On the other hand, some very perplexing cases are little-known.)

For our purposes, there are three important questions about a UFO report:

1. Was it reasonable to think that the report had "defence significance" (to use a favorite term of the British Ministry of Defence)?
2. If so, did the military and government authorities in possession of the report recognize this significance?
3. If they did, did they take appropriate action?

The modern UFO era began in the summer of 1947 with a wave of reports in the US. The first widely-publicized report was that of private pilot Kenneth Arnold near Mt. Rainier, Washington on July 24. (The term "flying saucer", by the way, resulted from a misinterpretation of Arnold's statement that the objects he saw flew like saucers skipped across water; they were crescent-shaped, not disk-shaped. However, the shapes most commonly reported during this wave, and for some time to come, were disks of one kind or another. The term "unidentified flying object", "UFO" for short, did not come into use until several years later.)

UFO reports have a strong tendency to come in waves, restricted in both space and time. We have no way of knowing whether an increase in the number of reports results from an increase in the number of UFOs (whatever they may be) in the skies; an increase in the probability that a given UFO will be seen by someone; an increase in the probability that a UFO, once seen, will be reported; an increase in copycat reports by others as people race to be second; or some combination of these four factors.

If we may be permitted to use the term "UFO" a bit loosely, there were several UFO waves in the half-century before 1947.

A wave of “airship” reports started in California in November 1896. By the time it ended, in May 1897, the “airships” had been reported as far east as Tennessee. Difficult as it may be to believe, journalistic ethics were even lower in the 1890s than they are today. Stories made up out of whole cloth were routinely published by “reputable” newspapers, not just by the likes of the *Weekly World News.* Many of the “airship” stories were newspaper hoaxes. Others were made up by the “witnesses”. Others were honest mistakes; more people knew Venus when they saw it in the 1890s than today, but the overly-excitable will always be with us. Almost everyone assumed that the “airships”, if they really existed, were the work of one or more American inventors, but it is quite certain that this was not the case.

There were many "landing" and “occupant” reports, but almost all of the “occupants” were said to be Earthmen, most of them Americans. There were a few reports of extra-terrestrials, but hardly anyone took them seriously.

There was a crash report. In April 1897, a spaceship hit Judge Proctor’s windmill, in Aurora, Texas, destroying the ship (and windmill) and killing the pilot, who was buried in the town cemetery. A hoax, of course--but when the story was rediscovered in 1966, many UFOlogists fell for it, and expended considerable effort, over a period of years, searching for the remains of the ship and the pilot.

And there was a cattle mutilation report. Also in April 1897, the crew of an airship lassoed a cow on a ranch near Leroy, Kansas, and flew away with her. Her hide, head, and legs were found the next day. Another hoax--but careless UFOlogists were still telling the story 80 years later.

Finally, there was a report of an unsuccessful abduction attempt by aliens, in November 1896, near Lodi, California. The government and the Army, understandably, ignored the whole thing; no one thought that any of the reports had “defence significance”.

There was another wave in 1909, this time in the Northeastern United States, Britain, New Zealand, and Australia. Other waves have not had such large gaps in the regions they covered! This time, some of the objects were described as airships, others as airplanes. Again, there were "landing" and "occupant" reports, but this time there were no reports of extra-terrestrials. In Britain, there was considerable speculation, some of it rather hysterical, that the "airships" came from Germany, which they did not. Government and military authorities kept their heads.

The next wave was in Britain from the fall of 1912 to the spring of 1913. The political situation had changed considerably since 1909, and these reports were taken quite seriously by the government and the military. The matter was discussed in Parliament, and the development of antiaircraft guns was accelerated. But, whatever they were, the objects were still not German airships.

Beginning in November 1944, air crews in the European theater began to report being followed by small balls of light, single or in groups. They were dubbed "foo fighters" by the Americans. The term came from the comic strip *Smokey Stover*, in which it was often said that "where there's foo, there's fire"; "foo" is an Americanization of the French *feu*, "fire".

The foo fighters were widely supposed to be German secret weapons, and, after the war, there were poorly-documented claims that this had been established. But they never gave any sign of hostility, and they were invisible to radar. On the other hand, the sightings did end with the fall of the Third Reich.

Allied Intelligence took the foo fighters quite seriously, although they did not succeed in learning anything useful about them.

Between February and October 1946, hundreds of reports of what were called "ghost rockets" were made in Scandinavia. Despite the term, by no means all the objects were described as looking like rockets, but many were. They were generally assumed to be Soviet missiles, despite the obvious fact that no one in his right mind would launch experimental missiles across foreign territory. Of course, we now know that in 1946 the USSR did not even have any long-range missiles, apart from a few captured V-2s.

The governments of Sweden, Denmark, Norway, Britain, and the United States took the ghost rockets very seriously indeed. Press censorship was imposed by Sweden, Denmark, and Norway, and the Swedish investigation of the reports was not declassified until 1983.

The US Air Force also took the "flying saucers" of 1947 very seriously. They were widely believed to be Soviet secret weapons--although some thought they might be *very* secret *American* weapons! The Air Force lost no time in organizing a project to investigate the UFOs, under the Air Technical Intelligence Center at Wright Field (now Wright-Patterson Air Force Base), near Dayton, Ohio. This was Project Sign; the name was undoubtedly significant. The project's existence was not classified, but its name was. Publicly, it was called Project Saucer.

The "secret weapon" theories rapidly lost ground to the "alien spaceship" theory. On January 7, 1948, just before Project Sign opened for business, Capt. Thomas Mantell, of the Kentucky Air National Guard, was killed while trying to intercept a UFO, it began to look as if the UFOs might be hostile. The public was told that Mantell's UFO was Venus, which was obviously preposterous; they were also told that it was a balloon. Thus began the "just say any old thing, as long as you identify it" tradition; this was to remain Air Force policy, except for one brief period.

In fact, Mantell's UFO was one of the Navy's Skyhook balloons. At the time, these balloons (manufactured by General Mills!) were secret, because they were to be used to carry cameras over the Soviet Union. After their existence was revealed, the public was told that they were used for scientific research--and some of them were. But Mantell did not know that such huge balloons existed. He greatly underestimated the balloon's altitude, and thought he might be able to intercept it in his P-51--which was not equipped with oxygen. He climbed too high, lost consciousness, and crashed.

In September 1948, Project Sign produced a Top Secret "Estimate of the Situation" which concluded that the UFOs were alien spaceships. It was rejected by the Air Force Chief of Staff, Gen. Hoyt S. Vandenberg, on the grounds that the evidence was insufficient. He ordered it declassified *and destroyed*! For years, the Air Force denied that it had ever existed.

One might think that Gen. Vandenberg would have ordered Sign to be expanded. He did not. It was quite clear to Sign personnel that he viewed *both* the secret-weapon theory and the spaceship theory with disfavor, and that he saw no need for more evidence. Sign promptly adopted an unequivocal "hoaxes, hallucinations, and misidentification of known objects" position, and serious investigation of reports stopped. In February 1949, the project was renamed Grudge, supposedly because the original name had been "compromised"; the new name was undoubtedly significant too. In December 1949, the Air Force announced, falsely, that the project was being terminated.

A special type of UFO, the "green fireball", was seen over New Mexico from December 1948 to December 1951. Lincoln la Paz, one of the world's leading meteoriticists, concluded that they were not meteors--for a number of reasons, not least the fact that one can hardly imagine an anomalous type of meteor confining itself to the skies of New Mexico for three years! Because of the concentration of highly-classified military facilities in New Mexico, the authorities were very concerned indeed.

In December 1949, the Air Force launched Project Twinkle, with the objective of collecting data on the green fireballs. Twinkle was entirely separate from Grudge, which was, of course, of no use at all. Unfortunately, Twinkle was under funded and incompetently run, and after two years it had accomplished nothing. (In fairness, the fact that the Korean War was getting under way at the time cannot have helped.)

In October 1951, Grudge was re-organized under Capt. Edward J. Ruppelt, and resumed doing its job. In March 1952, the project's name was changed again. It became Blue Book.

The summer of 1952 saw a huge wave that included the famous Washington National Airport incidents; on the nights of July 19-20 and July 26-27, there were multiple radar and visual reports of UFOs over Washington. In January 1953, the Central Intelligence Agency convened a panel of five prominent scientists, under physicist H. P. Robertson, to study the UFO problem. There seems to have been little or no concern in the CIA about the possibility of an alien invasion; the Agency was concerned, not about UFOs, but about UFO *reports*. The wave that had just occurred had overloaded intelligence agencies, and it it was feared that a future wave--perhaps faked by an enemy--could obscure preparations for an attack. Public interest in UFOs was also felt to have undesirable effects.

The Panel devoted only four days--short days, at that--to its task. It recommended that Blue Book treat UFOs as a public-relations problem, not an intelligence or scientific problem. The Air Force, not usually so ready to let the CIA dictate its policies, adopted this recommendation. Until its termination in December 1969, Blue Book devoted no serious effort to UFO investigation; even though there was a large wave in late 1957, and a large, if protracted, wave in 1965 and 1966. During most of this period, the Blue Book staff consisted of no more than three men and a secretary.

The justification for Blue Book's termination was provided by a study conducted at the University of Colorado, under physicist Edward U. Condon, between October 1966 and November 1968. It was widely criticized for dishonesty and incompetence.

Since Blue Book's termination, the Air Force and other US government agencies have, at least publicly, largely ignored UFO reports. Even when these reports were made by Air Force personnel, as in the case of the Rendlesham Forest incident at the Woodbridge and Lakenheath Royal Air Force bases in December 1980.

Although the UFO phenomenon is worldwide (there was, for example, a very large wave in France in 1954, when few reports were being made in the US), the only foreign governments that have devoted any substantial effort to UFO investigation are those of Britain and France. British investigations have largely been classified.

So why was this section included you might ask. The point of this section was to emphasize the political motivations and non scientific or military based approaches to them. Whether there was a real or perceived public “need to know” does not seem to matter for any of these cases. It appears that from a historical perspective that only political motivations have driven the investigations.

The bungled approaches from Sign, Grudge, and Blue Book suggest that either unqualified people led these efforts, or they were improperly handled, or they were a cover for some other effort (UFO or not). Whatever is the case, such fervor of conspiracy theories and government distrust was created as to create a permanent backlash for any future efforts. What real purpose could these projects have served as bungled as they were? They surely could not have prepared us for an imminent invasion. Such preparations would have to be performed in a much more methodical and scientific manner by highly qualified, diverse personnel.

5.2 Hide, Hail, Warn, Then Shoot – Maybe just Shoot First Anyway!

We can not tell you that ET exists or that he will visit us. However, the existence of many intelligent lifeforms within the universe is highly probable as shown earlier. If many ETs exist, it is reasonable to assume that one or more will visit the Earth at some time in the future. It is also logical to believe that one or more ETs could have already visited Earth. It is possible that we will be the ET at some time and visit one or more celestial bodies.

As the visitor or as the visited, we will need to know how to engage ET. We will need rules of engagement. Those rules will vary according to the category of ET. It would be a shame to treat a Benevolent ET as a Hostile ET. We would risk losing the benefits of that benevolence. It could be a tragedy to treat a Hostile ET as we would a Benevolent ET. We could lose our civilization or worse - our lives including those of the entire population of Earth. This is the fallacy of the SETI approach of welcome one and all ETs to Earth, especially without preparation. Which, if any, ET will heed SETI's invitation? We are unable to predict at this time. We would be better served shutting down SETI broadcasts and have them only listening for signs of extra-terrestrial intelligent life.

Our recommended philosophy is simple. **Prepare now, Survive later!** This is the best we can do. We continue to repeat this philosophy throughout this book with hopes to get the point across. To our knowledge, and as an educated guess, very little preparation is being made at present.

Someone or some group must be entrusted with the responsibility to make a decision as to the proper welcome in the event of an encounter with ET. That decision will be whether to defend against, negotiate with or submit to ET. That decision will be difficult as it will be mandatory properly categorize ET in order to offer the proper response to his approach. We need to develop techniques to assess the intentions and military capabilities of ET. We will be limited in doing so by our understanding of physics and the state of our technology development. However, we need to begin to think beyond our current understanding of physics and our current technologies.

Preparation for the hostile ET will be the most demanding case of the defined categories on ET's. This is true because so much will be riding on the outcome of the encounter. We should avoid the hostile ET if at all possible. If we look at the categories of ETs and their possible motives for visiting us, it appears that there is a greater probability of visits hostile ETs than from all other categories combined. As of today we would probably be wise to lay low and hope that he would go away; that is an unlikely scenario. Why would they travel the great distance required to get here only to sneak a peak and leave? What would their motivation have been to get here? Laying low would only work if he happened upon us by accident and had not detected SETI's beacons and signals. Laying low might also work if he had poor sensors to detect life. Or it might work if he had limited ability to land his craft.

We would certainly want to keep hostile ET off our soil. Keeping him in orbit may indeed cause him to use up precious resources which would diminish his ability to wage war against us. One indicator that ET is hostile could possibly be that he would travel on a large scale. Seeing a large force should immediately put us on the defensive and we should begin to consider the ET as a potentially hostile force. Obviously, he may attempt to travel in disguise. He would probably only do that if he were unsure of his relative military capability. We must develop significant military capability to ward off ET. One can argue that such weapons are a waste as there are no ETs and that they would take money from the traditional military. The fact is that any capability that we would build would probably have terrestrial application as well and would therefore not a be a waste of talent or money. We should make a show of our might in hopes of deterring at least some ETs who have limited confidence in their military strength. In the short term, we should build decoys in an attempt to make our military capabilities look greater than they actually are. We may want to adopt a philosophy of shoot first, shoot big and shoot often if we are certain that ET is hostile.

The benevolent ET will be the easiest category for which to prepare. We should be very happy to see the benevolent ET show up at our doorstep. He undoubtedly will bring technologies and techniques that will be quite helpful in the advancement of our society. Unfortunately, we are unable to forecast that category of ET with which we might first or later be forced to deal with. All we have to do with a benevolent ET is open our arms and our society and enjoy the benefits. Why if we knew only benevolent ET's would respond, we would encourage SETI to continue sending out its beacons and signals. In fact, we as a global society should spend tremendous sums of money helping SETI build new and better facilities to send out its signals. Since we don't know that only a benevolent ET will come to visit, we need to prepare for the worst-case visit and hope to enjoy the best case visit. We still need rules of engagement as much for our own benefit as any other reason. We will need technologies that we do not currently have to host benevolent ETs.

The other two categories of ET's produce no obvious unique challenges with respect to preparation for their arrival. Spectator or neutral ET's can be prepared for in the same manner as benevolent ETs. The preparation for researching ET's would require a combination of those activities for hostile and for benevolent ETs.

Regardless of the category of ET the most significant challenge is that of determining the ETs intentions and military capabilities. Regardless of the category we need to have well-defined leadership roles and interfaces for dealing with ETs. It should be well understood who has the authority to make commitments for and statements that represent the general populace. We need to develop and continually revisit a complex set of rules of engagement for ETs. We need to develop those technologies which provide us the greatest opportunity for success in the event of an encounter. We need to start now! **Prepare now, Survive later!**

From the above discussion we would suggest the best approach upon large-scale encounters is to hide from the ETs as long as possible while maturing our technologies. If they show up then we hail them in a friendly manner and warn them against hostilities. If the warning does not dissuade them, then we should adopt the philosophy that they mean us harm and we should shoot first before it is too late. If they have the technology to get here then they may be able to translate any communications signals we send them and be able to understand our hails and warnings. If they do not respond, then they are perhaps hiding something and so maybe we should start shooting immediately.

5.3 Dig In, Recon, Intel, Plan, & Act

From discussions with WWII vets about how the local populations survived during the Nazi invasion of their countries it is pointed out that their cities were designed much differently than ours. Most of their phone lines, power lines, water, gas, and other infrastructure components were underground. Most of the houses had underground shelters or basements. They were able to keep going even though their cities were rubble. Figure 5.1 is an actual photo from the time (Germany). The point is as a civilization we need to be prepared and "dug in" before we get attacked. Notice that the people in these photos are going about their daily activities. They are living underground surviving off of what little they could find.

Perhaps we will be faced with typical warfare on the second wave of any attack or

Figure 5.1. Actual photo of German city during WWII showing inhabitants survival which was mainly due to the fact that they were "dug in" before the war

war. If we are lucky enough to survive to a second wave, then standard battle theory may hold and we should approach the situation on with multiple front strategies, weapons, and tactics, as well as deploying typical second wave offensive capabilities. Unless our capability to make war was completely decimated in the first wave and we now must rely on the civilian population to wage an asymmetric war as discussed in Chapter 2.

A second wave tactic that may be quite useful in surviving an unfriendly alien counter is deception and denial, commonly known as D&D (not *Dungeons and Dragons*). Deception and denial are closely related with the distinction between the two sometimes being difficult to recognize. This tactic was also discussed in Section 2.5.

According to Webster's Dictionary it is "the act of deceiving; illusion; fraud." Deception is fairly easy to understand; it is simply trying to fool another being. It typically involves the use of visual or audio effects. It involves the use of such tactics as hiding tanks inside a cardboard structure. The cardboard structure is there to prevent an outsider from knowing the whereabouts of the tank. The use of decoys is another formal deception.

We've seen an old Western movie where one or two people would place rifles around a group to try to convince them that they were surrounded by a superior force and hopefully get them to surrender. Deception might also be hiding one's troop placements deep inside a mountain. Denial according to Webster is "a refusal of a request; refusal or reluctance to admit the truth of something." And denial implies false response to interrogation or inquisitive probes.

An example of deception might be to hide communications signal within a stream of communication signals. One might place a low-power transmitter for telephone communication beside a high-power television transmitter using a frequency very close to that of the television transmitter. One might use such RF techniques has spread spectrum and frequency hopping to deny the collection of one's signal. The use of encryption has been a well-known tactic for denying one's voice communications.

Obviously, there is no good reason to hide from an alien unless one is faced with an unfriendly or hostile encounter. However we cannot at this point project whether the first, or next encounter depending upon your point of view, will be friendly or unfriendly. While SETI broadcasts its signals to any and all available intelligent forms of life, we believe that the *Sixth Column* should be developing deception and denial technologies which will allow us to select what lifeforms receive our signals and have the ability to determine our military strength.

Of course we would welcome benevolent ET's or researching ET's to interact with us. However, on the other hand we would wish to avoid an interaction with hostile ET's; especially those with superior military capabilities. We should strive to survive and build and adapt our war machine to better face the ET threat. We should also evolve our defense from the survival tactics such as terrorism to more offensive and less defeatist type goals.

At a minimum, we should dig in deep. Probe and study the alien invader's capabilities. Formulate a plan to drive them away or defeat them outright. Then act on that plan following the principles of warfare discussed in Chapter 2.

5.4 Long Term Defense Plan

How did the Native Americans survive assimilation?

History can be a great teacher or it can send us down the wrong path. We can learn much for our preparation from the plight of the American Indian in the early United States. The white man was the Native American's ET. There are many parallels between the state of preparation of the Native American for their encounter with the Europeans and our state of preparation for an alien invasion. Some of those are:

- He did not know that the Europeans existed just as we do not know if ET exists
- He thought little, if at all, of an outside invasion;, we only think of an alien invasion in small groups
- He was making no preparations for the unknown; we are patiently waiting for the future to get here
- His technology was inferior to that of the unknown attacker; our technology will almost certainly be inferior to that of ET if he shows up on our horizon unless we begin to prepare diligently right away
- He was making no efforts to improve his technology base to better defend himself; we are making no efforts to improve our technology base beyond that of conventional weapons for fighting Earth based conflicts even thought we could simultaneously improve those as well
- He, like SETI, welcomed the aliens without regard to intent or military might
- He was disorganized as a lot of little bands who each had limited military strength; we are also a group of small nations each with its own agenda and limited military strength – forget how the US stacks up against the rest of the Globe
- He was a fearless warrior because he believed in himself and his skills; we have fearless warriors who have distinguished themselves in many battles and who believe in themselves and their comrads
- He was doomed by his limited firepower because he waited too long to adapt the technology of the white man; we may not have time to adapt ET's firepower so we should be developing advanced weapons while there is time
- He had faith that the "Great Spirit" would protect him; we believe our creator will deliver us but perhaps he expects us to contribute to our welfare and longevity
- He thought he was a much more powerful warrior than was actually the case; we think we are better warriors than we are because we are accustomed to fighting weaker forces – ET is probably not going to be weaker than we are today – we probably will not be able to establish air superiority with our current capabilities

- He was without a plan for how to address an outside threat; we do not have a plan for dealing with a significant outside threat
- He was unprepared; he did not practice **Prepare Now, Survive Later!**; we are not **Prepared Now to Survive Later!** It is Time to prepare now!

In each of those cases we are like the American Indian. Unlike the American Indian, if banished from our tribal lands, we have no place to go and no way to get there. We should learn from the American Indian's plight and **start preparing.**

It has been wisely said that those who fail to learn from history are doomed to repeat its mistakes. We have a chance to prevent some of those mistakes; we need to get started right away. Otherwise, we will be forced into terrorist tactics and a never ending "underground" resistance which in the long run might wear down and prove ineffective. And in turn, the long occupation might be the end of humanity.

The better approach is to build the biggest and most effective war machine we can manage and to follow the philosophy that it is better to have a gun and not need it than to need a gun and not have it. We will need big guns and lots of them if a hostile ET knocks at our door one morning as our wakeup call.

5.5 Did We Win?

If we are attacked by hostile ETs there are a variety of outcomes possible. Some of these are:

- Victorious (not likely)
- No Survivors
- A Few Survivors Leading to Species Extinction
- Frontier Species Survival (rough)
- Cohabitation (mutual benefit, as slaves, as food)
- Evicted (where do we go and how do we get there)

A harsh outcome of being conquered could be that we are evicted from Earth. This might be because our conqueror wants Earth for its food growing ability, its natural resources, its environment or other purpose. We would be displaced much as the American Indian was from his native lands. Where will our reservation be? How will we get there? Unlike the American Indian we will not be able to walk to the reservation; but it will be a long and difficult trip.

As the conquering ET will almost certainly not be human, we should not expect humane treatment; that will probably be a foreign concept to him. We can demand humane treatment; however, we will be absent a compelling posture in our argument.

ET's view of humane will most likely differ significantly from ours. We also will have limited negotiating power. If we are lucky our conqueror could be persuaded to transport some portion of our populace to another location in the universe. It almost certainly will be far less accommodating than Earth as it will most likely be void of the infrastructure that we have become accustomed to living in and around. If it were better or equal to Earth, why would the ET not just go there and meet little or no resistance.

Would we have to carry the military might to take over another planet? Why wouldn't ET just take that planet and save his resources? If ET were to transport at least a portion of the populace it would probably be to satisfy a future need for slaves or food. We would be much more vulnerable to his future attacks.

Unless the conqueror has a huge fleet of vehicles, only a small portion of the Earth's population could be taken to the new reservation. It would be a tremendous logistics problem to transport several billion people even a distance as short as to Mars. The trip would probably be something akin to the Bataan Death March of World War II.

Who is to decide who gets transported and who stays? Who determines the order? Who determines the fate of those beyond the transportable population limit? Should each of us negotiate on our own? Is it true that you get what you negotiate? Would there be life boats in the sky? Who would get a seat; the women and children? If it were the women and children, could they sustain themselves on a new planet; perhaps chivalry would result in a cruel ending far from home.

5.6 Chapter Summary

This chapter covered what we would consider the response to an ET invasion. In Section 5.1 we gave examples from projects Sign, Grudge, Blue Book and others as how not to handle such things. The known public information about the UFO investigation programs conducted by the U.S. government in the past were not motivated by a best response to the potential ET threat rather than some sort of public appeasement agenda. Whatever the agenda of those programs, they were not handled in a way that seems best for a real alien threat.

Sections 5.2 and 5.3 discuss possible large scale actions in case of an invasion. We might want to be friendly and say, "Hi how are you?" And if we get no response it might be time to start shooting. If they could travel from there to here they would understand how to respond to our hail. After the initial onslaught of an invasion we should conduct intelligence gathering missions and plan for a more concise attack against the invaders.

Sections 5.4 and 5.5 discuss the possible long term defense plans and the outcome of an invasion. If our defense plan is not short sighted we might survive. The question becomes, what type of survival would that be.

6

The Sixth Column – Somebody Should be Preparing

6.1 The Necessity of Preparation

Of course there is no guarantee that an alien invasion will occur in spite of statistics. In fact, we should probably hope that it does not occur. But inevitably, something will happen be it good or bad. It may be a benevolent civilization that brings us all types of technical and medical advances as well as the latest copy of the *Encyclopedia Galactica*. That would be wonderful. Or they may just let us continue on our way without interfering. They may bring us great prosperity by setting up trade agreements with us. Those situations too would fine and preferable.

But what if a hostile invasion does happen? We had better be prepared in order to avoid shear and total panic and chaos that would doom us to certain defeat. Even if the invasion is benevolent, we should be prepared in order to avoid losing that opportunity, or worse, turning it into a hostile encounter. Proper preparation offers our best chance for survival in the event of an alien visit be they friend or foe or just trying to determine which we are.

The need for preparation is pointed out by a recent conversation with a young married couple. The couple involved knows one of the authors of this book. They knew something of the writing of the book. They questioned the probability and likelihood of an invasion. Upon hearing the author's response, the young husband immediately espoused his belief that there is no such thing as aliens. This belief may serve him well throughout his lifetime. If the Earth is not invaded during his lifetime, for all practical purposes he was right. The young wife on the other hand was terrified that an invasion might occur. In panic she questioned, "What are we going to do if they do come?" She was quite concerned that there be some plan to handle the invasion. We agree. She was absolutely correct! There must be a plan in place for us to have the best chance of a successful encounter. A successful encounter is one where our civilization is at least preserved, if not advanced.

It has been well documented that lack of planning was a serious setback for NASA after the Challenger catastrophe. NASA had no plan and no one to step forward and accept leadership. Statistical failure analysis conducted by the U.S.A.F. indicated that there would be a major accident in every 30-40 launches. NASA was familiar with those statistics, in fact paid for them, but took no action to prepare or the inevitable accident that befell Challenger. While NASA did not know that Challenger would have the accident, it knew one was coming and failed to prepare. A month or so of chaos followed the Challenger explosion whose launch was within the statistics. The President ultimately had to appoint a commission of celebrities who were ill equipped to handle a technical investigation of this nature. But a high profile group was needed to reassure the populace. There is room to debate the findings of that group because of their lack of scientific training and skills. In the event of a friendly alien encounter, a month or so of chaos and lack of direction could cause it to become a warring situation. A one month or so period of chaos in the event of a hostile alien invasion resulting in an eat-or-be-eaten situation could lead to being eaten before a resistance could be mounted; a resistance might never materialize. NASA has shown us the wrong way to handle an emergency situation. But NASA is not the only government entity that made these lack of preparedness mistakes.

When ET steps off the ship and says "Take me to your leader", we should have a leader designated and ready to entertain them. ***Let's get it right in the event of an alien invasion; let's have a plan***!

We must prepare as a nation. This is true even though the strength of all the Earth's nations maybe required in order to have any hope of repelling the advance of the most powerful hostile aliens. It is lamentable, but there are a number of barriers to preparing as a global entity. Among those are the following:

- Nations must be willing to disclose the details of their technical systems. Nations such as the US, France, Great Britain and Russia would be disclosing to countries such as Iran, Iraq, North Korea and Colombia who could be secretly trying to replicate those capabilities. Data acquiring countries would learn how to counter and deny those systems doing their intended missions within the borders of their countries. Rogue nations would be willing to join such an alliance just to gain insight into other countries technology base.
- The sharing of technology would render expensive sophisticated systems marginally useful at best. More likely, they would become useless. This would lead to a technology retrenchment, not just stagnation, in democracies as governing bodies would be forced to stop spending on useless, high tech systems. All the better for ET to eat you with.
- Dictatorial rogue states would be in a far better position to invade and conquer their hapless, cooperating neighbors.
- Does this global alliance jointly develop the technology to fight hostile ET's and in the process explore the universe. It follows logic that to defend one's self from an adversary, one needs to know the capabilities of that adversary. Ideally, one builds a better defense and offense than the adversary. This is the ethic of sports;

to win a championship one builds a stronger, more athletic team. We can and should also learn from sports that one must continuously upgrade one's abilities to remain the champion. Be the champion one year and be the most improved the next to sustain one's position.

- Who would pay for such an alliance? There are countries in the UN today, which make little or no contribution.
- When the chips are down, whom can you really trust? We have recent examples of countries breaking their bond with other countries. There is North Korea and its violation of a nuclear capability accord. There is Iraq and its previous use of weapons of mass destruction. It may sound callous, but perhaps such events are actually good for the global neighborhood. This is particularly true in the areas of diplomacy and technology development. As is true for the consumer, a little competition is a wonderful thing. Greed is still going to be with us. Should an invasion occur, there will be one or more countries which bolt from the alliance in an attempt to negotiate a better deal for themselves and, ideally, their country. It is difficult to fault them for trying to obtain the best deal for their citizens; that is after all the job of leaders – the common good of their constituency. However, it is easy to condemn them for personal greed.
- Who would be in charge? Is it the largest contributor or the best athlete? Unlike the United Nations, the leader needs to have technical, military and diplomatic skills or a small but talented set of advisors in those areas. Truth is that there is probably no one person with World (truly World as in Global) class skills in all those areas. Therefore, advisors would be required. They must be the globe's leading authorities in their respective fields, not political appointees. What if the leader ignored the advice of the advisors?
- The United Nations is exactly where the leadership should not be because of the veto and blockage rights of member nations.

There are of course counter arguments that would lead to setting up a global controlling entity. Among those are:

- Am I my brother's keeper? What are the implied and stated obligations of the US as well as other leading countries to our less fortunate neighbors?
- The combined strengths of the globe or a large portion of it.
- What is the sum of the capabilities of the global community versus those of the US?
- The beauty contest answer, "World Peace." What are the implications to global peace and tranquility? As Utopian as "World Peace" sounds, it is probably a bad idea. Peace and harmony on the globe would lead to technical retrenchment; there would be no motivation for Governmental funding of advanced technical reach. Immediate profit would be the only imperative for research and development. The Captains of Industry have shown little interest in funding anything without

immediate payoff; the incentive plan for the Captains of Industry and the stock valuation approach would have to be changed to get industry participation in the form of investment. Imagination would be put on hold unless the global community provides funding for research and development. Then there would be the fight over where the monies are spent. There would certainly be no motivation for companies to invest in offensive or defensive weapons, as there would be no immediately identifiable market for their wares.

These arguments and other arguments should be carefully dissected and considered. The authors believe that a national entity will survive when the proper scrutiny is applied to the arguments. We offer our suggestions on how to organize such an entity.

A small, special agency should be established to lead strategy development. It must have special status. Therefore, it must be separate from the existing politicized Government agencies. It must be highly secret like the Citadel in Robert Heinlein's book *The Sixth Column*.

This group would have better direction and leadership than Heinlein's Citadel. Access would initially be limited to a dozen or so world-class scientists, a leader with a diverse technical background and leadership experience dealing with technical wizards, a few members of the Administration and the few members of the overseeing Congressional Subcommittee. Security access would initially be limited to less than 75 persons. Security access would be for one's lifetime. Information flow would be restricted to within that select group on a need-to-know basis. Breach of security to disclose the existence of any detail about the project would be deemed treason.

The leader should be selected first. This will be the most difficult task in making this venture a success. The leader would at best not be a well-known or high profile person. A high profile person would be difficult to hide from public view. The media would soon be attempting to uncover a high-profile person. Finding a low-profile person or any person for that matter, with the proper talent to execute this role will be difficult. It will be a demanding process; especially, since the nature and propose of the search must be protected from those discussed and possibly interviewed.

The President's science advisor and the Secretary of Defense should recommend the person to fill this nonpolitical position without political interferences. This appointee would be approved by the SSCI. The leader would in turn select the technical staff. The movement of the leader and technical staff must be done in such a manner as to avoid arousing suspicion. This team would perform the technical planning for exploration as well as defense readiness. They would also develop the strategy for dealing with an alien encounter on Earth or during our galactic explorations. The plans and strategy would require updates every other year in order to remain aggressive. Leadership of the team would have to change periodically with each new administration. The old leader is shifted to an advisory board within the organization.

The technical wizards would come from industry and academia. They would be composed of physicists, chemists, biologists, engineers, mathematicians and psychologists. They would be collocated in a secret location far outside the Washington area and its long shadow in a well secluded undisclosed location. These people should be paid above the industry average with annual raises also above industry averages, as they will be committing their lives and careers to this project. These people will be taken from the normal career progression and the fight to climb the corporate and academic ladders to acquire riches. They will lose the opportunity to gain professional notoriety.

After the group has had a year or two to organize its thoughts and develop a plan, the group would be expanded slightly and gradually in areas identified as critical to success. As planning continues, it will be necessary to bring in support for research and development from industry. Even as expansion occurs, there should never be more than 5,000-10,000 persons granted security access at any given time.

The development of a national team with allegiance only to the project would be assembled from industry at-large. The staff would remain with their respective companies and the companies would be compensated for supplying talent. The staff would all be committed to a career of service to this cause. The staff would not be allowed to leave and seek riches as a result of their exposure. Should they ever leave the staff, they would not be allowed to return for any reason except a personal ET encounter. Corporate executives should not be accessed to the existence of the organization or to its mission. They should only be told that support is needed for a national initiative. The facilities should be Government financed and owned.

This *Sixth Column* type organization must have well-defined roles, responsibilities and authorities. All of their efforts must lead to the development of strategies and technologies for an alien encounter or for galactic travel by Americans. In order to function properly the organization must have autonomy as discussed earlier. It must also have adequate budget and absolute control of the expenditure of the budget to meet its mission. It is imperative that this organization be put in place in such a way that there is no need for personnel with previous government and bureaucratic experience. Although a liaison group to that community might prove useful.

The organization should have broad roles, responsibilities and authorities. This is necessary, as it must build itself based upon a variety of disciplines. Chief among those roles, responsibilities and authorities are:

- development of galactic exploration technology
- development of ET engagement tactics, strategies, and plans
- talent recruitment and retention
- advanced weapons development
- exploration planning (history, archaeological, theological, astronomical, etc.)
- development of necessary linguistics tools
- potential and real encounter investigations (what Blue Book should have done)
- encounter management (*Men In Black*)
- compiling and maintaining data on all science and engineering breakthroughs world wide (including foreign classified breakthroughs)
- reporting to appropriate individuals within the Administration and Congress

- others as determined and assigned by proper oversight authorities.

Each of these is a serious and necessary function. These will be addressed individually.

Preparation for galactic travel will span many disciplines. The organization will have to spend money to develop technologies that can make that travel possible within a reasonable travel time frame. It will be necessary to undertake these developments outside the realm "Normal Science". Waiting for anomalies will take to long for the necessary technologies to be discovered. The organization must be proactive. It must not simply think out of the box; it must forget that there is a box at all. Thinking outside the box still implies classical restraints. We must forget the concept of a box in order to accelerate development times.

It is highly likely that a return on investment for such technical development would be tremendous. We have enjoyed the economic and sociological benefits of spin offs from space and military technology such as microwave ovens, high-definition television, home computers and wireless communication. It's possible that the spin offs from galactic travel technologies would have greater economic and lifestyle impacts.

The organization will have to procure and manage the production of the devices required for galactic travel. There are at least three technical areas that will need significant effort. Those are:

1) high-speed propulsion technologies,
2) accommodations related technologies for the travelers and
3) training of potential travelers.

Accommodations for galactic travel will have to be different from those for shuttle and space station astronauts. Those devices and technologies must be identified so that procurement can take place. Undoubtedly there is a lot to be learned from Astronaut, Navy Seal and other military Special Forces training and will prove to be beneficial. However there will be new challenges facing these travelers. Those must be anticipated and a training program developed which is unique to these individuals and those things that they will face.

Strategies must be developed for an encounter with ET. The strategy for an encounter on Earth will have to be different from the strategy for an encounter on ET's home planet. The organization must busy itself with development each. The strategy for an encounter must account for both benevolent and hostile situations. It must decide who will be in charge and each event. It must detail the steps required to execute the strategy. And it must have strategy off ramps leading to an alternative strategy if the leading strategy is assessed as being inadequate. ET may never come; however, the preparations will be well worthwhile.

The availability of the proper talent is always critical to the success of the mission. Unlike the classical military tactic of selected best all around athlete, successful encounters with ET will require the best talent for each function. This mission will have the best chance of succeeding by, for example, having the best electrical engineer for an electrical engineering position. The organization might have the best mathematician for doing the mathematical calculations that are required. It will have the best psychologist in the role of psychologist. And so on and so on. However, past programs, companies, and organizations have shown that a few generalists or all around athletes will help tie the developments of the specialists into realizable systems and technologies. Actually, such an organization must obtain the best of both worlds, so to speak.

Obtaining talent will be difficult. Retaining talent will be more challenging. It is critical that management focus its efforts on a daily basis to keeping the talent pool challenged and satisfied. It is too important to mission success that human resources and retention not become an issue.

Very talented individuals are not easy to find. If these talented individuals become dissatisfied with the effort for some reason it would in most cases prove more beneficial to address the problem with the individual than to take the time to train perhaps less capable individuals for replacement. Issues within the program such as consistency and security (among many) would arise. These are of course management issues that seem obvious. However, standard industry and government management protocols need to be streamlined to give the organization proper internal authority and flexibility as well as efficiency that would be required for mission success.

In the event of the hostile encounter it will be necessary to have weapons, which can succeed against ET. Classical military weapons are discussed in previous sections of this book. Some of them, perhaps even all of those weapons, will be effective against aliens. However, we should again be thinking outside the realm of "normal science" to develop much more capable and devastating weapons. If a hostile ET arrives, we may need a vast menu of weapons to fend off that encounter. It might be that obtaining these weapons is immediately beyond our realm of understanding and therefore we must look outside of our society for them. This is the premise of the *Stargate SG-1* television series.

Planning for galactic exploration should begin now, as there is a chance that we will run out of selected natural resources, which could possibly limit the population and longevity of earthlings. In such a case the only likely scenario for replacing those resources is via visitation to or from other celestial bodies. We must either set up a trade network with one or more other galactic civilizations, harvest those resources from an uninhabited celestial body or take those resources from other galactic civilizations by force.

In each of those scenarios, travel well beyond the Moon appears to be required. We may choose later to do such exploration only because we can or simply because we have a desire to do so. A great deal of planning will be required for such an event. A good science fiction example to consider in detail here is S*targate SG-1* whereas traveling through the stargates across the galaxy to find technologies to defeat an impending invasion is the only thing that saved humanity in every season finale over about nine seasons.

The never ending battle with Congress to continue funding the Stargate Program on that show is quite probably an example of what would happen in real life. The House and

Senate committees were continuously trying to cut the funds to the program, which was always "claiming" continuous impending doom. Expect the same from the real life counterparts. A contingent of the *Sixth Column* should be prepared always for this problem.

On Earth linguistics are at the heart of communications. In the event of encounter with ET, we undoubtedly will attempt use linguistics to communicate. As mentioned elsewhere in the text linguistics development has failed to keep pace with technical developments over the last few centuries. This organization could and should be responsible for advancing linguistics. Such advancement would ideally be beneficial in an encounter with ET. There would certainly be significant spin-offs that would be helpful to the classical military in the area of encryption and decoding encryption. One such spin off technology might be the famed "universal translator" as mentioned in so many science fiction works.

Any credible report of a potential encounter with ET must be investigated. That investigation should be conducted by the organization that has a procedure and is skilled at the execution of that procedure. The *Sixth Column* organization must develop that procedure. It must be very detailed, including essential dos and don'ts for investigators. The organization must be very agile and yield to respond immediately to all credible reports of encounters. This organization will not become another *Project Blue Book.* None of the results of any findings will be made public. As mentioned elsewhere in this text, there should also be zero public knowledge of the *Sixth Column* organization. So, encounter investigations will be kept under the utmost security.

If a real encounter occurs, we must be prepared. Since the organization will already be on the site investigating the incident, it is logical that they continue by providing encounter management. The organization that has been developing technologies and strategies for ET encounters will certainly be best prepared to manage such an encounter.

As is true of any government agency, it will be necessary for this organization to communicate its plans and accomplishments to the Administration and to the necessary Congressional committees. Such communication should be from a remote site. It would take advantage of television network conferencing and secured telephones. The staff of the organization should avoid frequent or periodic trips to Washington in order to maintain the secrecy of staff identification and project location. It is possible that a liaison to Washington could have a completely unrelated public title and reason to periodically and often visit the Capitol. However, general organization staff staying out of Washington also will minimize the tendency to drift toward bureaucracy and politicization. This organization must be separate, autonomous and special to succeed. It must be autonomous and special because its mission is unique and perhaps critical to longevity of the human race.

6.2 The Need for Secrecy of the *Sixth Column*

So why would we need all the secrecy? In this section we offer reasoning for the required secrecy of the *Sixth Column* organization. As mentioned previously there are many factors that suggest that the organization could not exist in the public eye with the current political and diplomatic status of the world. It is very likely that no other countries would wish to share their best and most advanced military strategies and technologies with any other countries.

Most likely, the US does not typically tell even NATO allies its deepest secrets. The technical advances required to defend the Earth against an ET invasion are paramount and would absolutely be the most advanced technologies ever created on Earth by any government. It is not practical to believe that any government would want to share that information. There are many reasons for this. Global security and stability of governments could be compromised if the wrong rogue nations were suddenly given access to weapons much more powerful than any previous WMD.

The dissemination of these Super WMDs could spark a new Cold War. Or, terrorists might gain more access to WMD technologies if they become more globally widespread. From this point of view, the sharing of Super WMD technologies is unwise.

Provided that the *Sixth Column* maintains its nonexistent status, the above-mentioned Cold War and terrorist activities might be avoided. On the other hand, if other governments discovered the existence of the organization it might be perceived as treaty violations and would spark a new arms race and Cold War. So, politically speaking, absolute secrecy is required.

One might ask why we simply do not share the information of the reasoning behind the *Sixth Column's* existence with the world and request it be an international organization. It is difficult to convince ourselves that this possible threat exists. How many Americans do not understand, believe, or care that the possibility of ET invasion exists? Whatever the number is, it will be much larger in lesser-educated countries of the world. It would be very difficult to convince the world to spend the money and resources on such a perceived to be outlandish project.

What if the international public could be convinced? A worst-case scenario would be that the public panicked and mass hysteria were created worldwide. We are not big fans of the mass hysteria concept. The entertainment community has desensitized the general public to the concept of alien invasion for nearly a century and they are probably ready. Some may panic, but not on massive global scales – perhaps, citywide scales. But then again it did not take much to cause a riot in Los Angeles following the outcome of the O.J. Simpson trial. What would happen following the release of information that alien invasion was impending? Consider the "planned" and "organized" riots during and after Hurricane Katrina in New Orleans in 2005 as a real life example.

More likely of a scenario would be either indifference or too much interest. If the world were finally convinced that ET invasion was possible, the press would tend toward claiming it either inevitable or impossible, but at least a waist of money. The general public tends to believe the press and can be swayed by them. In the US this could lead to either the death of funding or the tremendous growth of the project depending on which side the press takes.

The risk of the project being shut down due to lack of public support should be zero. But in the public eye it would be finite, much larger, and depending on public opinion polls of the day. Also, too much public interest could lead to so much unneeded help that the project could get bogged down with public affairs. The only way to ensure neither of these scenarios occurs is to take the public out of the picture. If the human race is to survive, the *Sixth Column* organization must be created and maintained at high funding levels. And it must be kept out of reach of the public opinion polls.

Also, if the organization were made international and public it would become politicized even more so. Various countries would make political and economic pleas to the United Nations to give them tasks under the program. In that case, the right people for the job may not be given the job. The right political answer may outweigh the right technical answer. This is a situation that is intolerable for such a serious concern as the survival of the human race.

This same argument applies to it being a purely American public program. The organization could fall to the same problems that NASA has in that there are multiple NASA centers in multiple congressional districts. In most cases, whether NASA will admit this or not, the political need for a program within a given congressional district tends to have more weight than a technical need. Some centers are given center mission definitions that will not allow them to create or develop anything real so as not to overstep the center mission goals of a different center.

This is the way the so-called NASA Technology Development programs work. As of 2003 certain centers were given absolute mandates NOT to develop a technology far enough to conduct flight tests because that is another center's mission. The reason for this is not because one center or the other is more equipped to handle further development of a technology. It is due to the internal NASA policy, politics, and protocols. As of 2005 it is difficult to tell that any of this has changed or ever will.

Of course this happens with other government agencies as well. The infighting between the NRO and CIA runs rampant throughout their histories. We suggest the reader check out *The Wizards of Langley* by Jeffrey T. Richelson for real life examples of the infighting between these organizations.

Although, many of the bugs within their management system have been ironed out some still exist. The Army, Navy, Air Force, and Marines have had never ending battles about whose mission is whose. Politics always seems to rear its head in these battles. It should never become a question as to what entity or groups can solve what problem. If a smart guy or group has a solution to a problem, no matter what their mission charter is the solution should not be ignored for political reasons.

So, the *Sixth Column* must remain as one autonomous group whereas no internal political squabbling and infighting is allowed or tolerated for more than a microsecond. A single Director *Sixth Column* (DSC) must be appointed and must have total managerial approval or disproval of all organization activities.

Of course the DSC should have deputies with delegated authorities and various science and technical advisors, but the DSC must have the last word. The DSC should answer to the President of the US only not to any committees. Of course, a few committees such as the House Permanent Select Committee on Intelligence (HPSCI) or the Senate Select Committee on Intelligence (SSCI) might be involved in budget planning. Also, interfaces through the President to the Joint Chiefs of Staff might be useful for strategic and logistics planning. However, the budget line should be permanent and untouchable by political action and natural inflation growth should be added yearly to the line at a minimum.

Another possibility for the organization's funding might be to allow it to earn profit through its technical advances. Technologies could be seeded into the public via dummy corporations that in turn could earn profit and pay royalties to the *Sixth Column*. The organization's mission is very important and no doubt will be very very expensive. Royalties from technology seeds might prove beneficial. This is similar to the *Men In Black* scenario.

Another possibility is to let the *Sixth Column* strategist publish possible scenarios as science fiction novels. Feedback from "fandom" will be free input and a technical sanity check. And there could be funding earned by the organization. A publishing firm could be set up and owned by the organization (it might even be called Sixth Column Press). The authors could be paid the royalties and the *Sixth Column* earns the publisher profits. The books could be made instant bestsellers by using a small amount of the research budget to buy 50,000 copies of each SCP book. In paperback at about $8 apiece, one bestseller would cost the organization only $320,000 dollars. The organization could publish fifty books a year and guarantee that a fifth of them would be bestsellers for only $3.2 million. The profits on them would actually pay for themselves in such a scenario. And achieving bestseller status would improve the chances of increasing the audience of the book – book buyers like to buy books that are bestsellers. Once a few bestsellers are created the company could promote those authors and new authors more effectively. The publishing company would achieve at least a few things: profit, widespread critique of the concepts, and public desensitivity to those concepts.

Another reason for very deep security is that it will help ensure the survivability of the organization in the event of an ET invasion. If the facilities are obscured as much as possible from detection from the public, perhaps they will be obscured from detection by the ETs as well. It might be impossible to keep the organization facilities permanently undetectable from the ETs, however the longer it remains a secret from the ETs the more time we would have to study and prepare a counterattack against them. As the organization grows, it should be spread into multiple physical locations with various backup facilities available in the event that some of them are destroyed. A compartmented cell structure, a neural net, and a hub system such as the Internet are good examples of how this type of arrangement might work. Major portions of the Internet can be destroyed but communications through it simply route around the damaged areas and continue on to their final destination. Various *Sixth Column* hubs (facilities) will enable a robust organization that might have a better chance of surviving an ET invasion. The hubs should be located globally not just across the country. And perhaps, there should be hubs in orbit, on the Moon, and elsewhere.

6.3 Organizational Structure

The DSC who answers directly to the President of the United States heads the *Sixth Column* organization. The DSC can brief the HPSCI, SSCI, and Joint Chiefs but only after the approval of the President. The Sixth Column must be special. To make it special it must have special status. It must be separate and protected. It must remain outside the political fray. It should be studied in great detail if the HPSCI and SSCI should be completely briefed on the *Sixth Column's* actual mission. It might serve security better to only partially brief these entities.

Full disclosure to individuals that do not require background checks is risking leaks. The only way that the *Sixth Column* can exist, as discussed previously, is if it remains completely secret from the rest of the world governments. The risk of leaks versus the need for having a fully briefed HPSCI and SSCI should be determined at the onset of the *Sixth Column's* birth.

There is also the possibility that the *Sixth Column* should report to the HPSCI and SSCI so that the Executive branch of government is not the only one aware of what it is doing. Although the DSC will only take orders from the President he could offer a monthly or yearly status brief to the HPSCI and SSCI. These briefings would enable a check and balance to be in place, which is a major cornerstone of the American government's success and probably should not be left out of any American organization.

By law the existence of the *Sixth Column* must be protected; no matter how much or little interest the President has in the organization it must remain. Also, the intelligence committees cannot remove funding to the organization below predetermined baselines. The line item for the *Sixth Column* must always be in place to at least a minimum level. Negotiations and briefings between the President, HPSCI, SSCI, and DSC should be used to determine yearly budget increases or reductions.

Below the DSC is the DSC's staff which includes the Chief of Security, Legal, Administration & Budget, Logistics & Support, and the Chief Scientist. Also, the DSC has six Deputy DSCs (DDSC), which answer to him. The DDSCs are in charge of the six different Directorates of the *Sixth Column*. A detailed organization chart is given in Figure 6.1 and shows the various Directorates as well as Divisions within each Directorate.

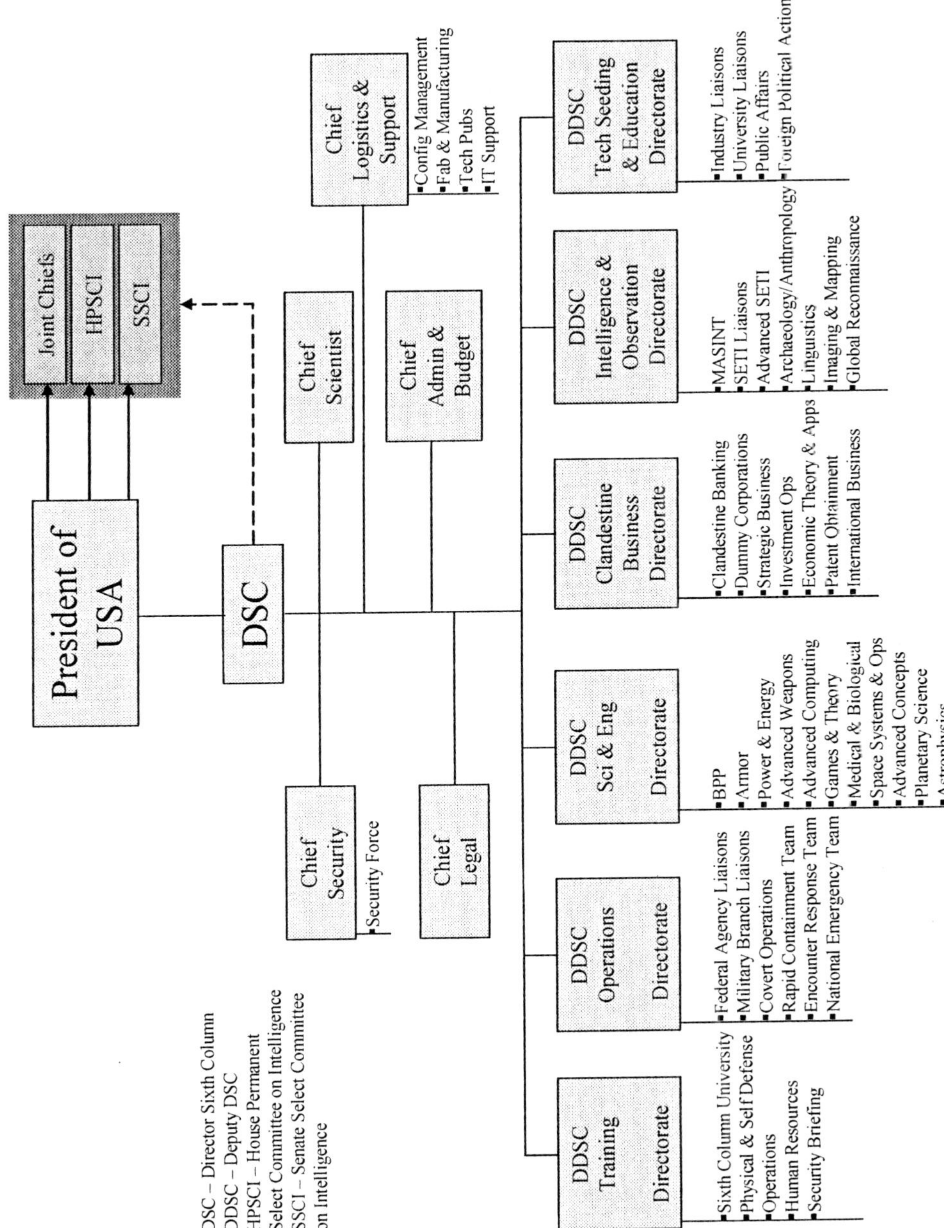

Figure 6.1 Organization Chart of Sixth Column

6.4 Organizational Rough Order of Magnitude Yearly Budget

The organization chart shown in Figure 6.1 shows a detailed organization for the *Sixth Column* that would ultimately consist of as many as three hundred employees, maybe more. The employees would all be highly qualified and therefore must be highly compensated in order to maintain job satisfaction and avoid high attrition rates. The salary cost of the organization is miniscule however when compared to the infrastructure requirements.

Assuming three hundred employees at a burdened (full cost accounting model) cost of $400K per year each, that is only $120M per year for very high salaries and overhead. This is not an exceptionally large amount of money when considering the cost of typical large-scale government programs. In fact, this is a relatively small amount of money when compared to the defense budget or even the NASA budget (which is very small compared to DoD). Eventually, the organization might grow to be as large as three thousand employees or more.

The infrastructure (buildings, laboratories, cafeterias, computer networks, security systems, etc.) will be a considerable part of the budget for the organization. We will assume that there are three separately located facilities each equipped with runways and launch facilities for at least Delta IV rockets. The Kodiak Launch Facility in Alaska was constructed for about $40M in 1998. We will assume that in 2005 the same facilities could be constructed for about $50M. Also, hiding most of the facility underground will add to the cost as well.

However, there are many ICBM underground facilities presently for sale for only tens of millions of dollars. It should be investigated if some of these old facilities could be refurbished or if it would be better to start from scratch. At any rate, initial costs for each underground facility should be less than $100M at worst case. Therefore, initial infrastructure cost is about $300M. We will assume that following year infrastructure upkeep is twenty percent of the initial startup cost. Therefore a yearly upkeep budget of $60M must be allowed.

The largest cost accrued by the *Sixth Column* would be its research efforts. Advanced physics, engineering, space, biology, and many other types of research would need to be funded. Figure 6.2 shows a page from the DoD Fiscal 2003 budget report showing the research budget for the Defense Advanced Research Projects Agency (DARPA) for the years 2001-2003. The total budget for the DARPA research was a little less than $3B per year. These DARPA programs are similar to efforts that the *Sixth Column* would need to undertake. However, DARPA programs tend to attempt technology advancements by an order of magnitude or so. The *Sixth Column* will need to trigger technology advancements by many orders of magnitude. Also, DARPA does not take technologies to fieldable systems. Typically DARPA will fund the basic research and then hand the program off to a different branch of the government or to an industry partner.

UNCLASSIFIED

Defense Adv Research Projects Agcy
FY 2003 RDT&E PROGRAM

EXHIBIT R-1

APPROPRIATION: 0400D Research, Development, Test & Eval, Defwide

Date: FEB 2002

Line No	Program Element Number	Item	Act	FY 2001 (Thousands of Dollars)	FY 2002 (Thousands of Dollars)	FY 2003 (Thousands of Dollars)	S E C
2	0601101E	Defense Research Sciences	1	99,647	142,303	275,646	U
		Basic Research		**99,647**	**142,303**	**275,646**	
9	0602110E	Next Generation Internet	2	14,461			U
14	0602301E	Computing Systems and Communications Technology	2	310,496	355,494	424,940	U
15	0602302E	Embedded Software and Pervasive Computing	2	47,876	65,561	80,000	U
16	0602383E	Biological Warfare Defense	2	166,224	146,688	173,000	U
18	0602702E	Tactical Technology	2	210,790	164,056	180,392	U
19	0602708E	Integrated Command and Control Technology	2	39,565			U
20	0602712E	Materials and Electronics Technology	2	255,626	344,554	440,550	U
		Applied Research		**1,024,430**	**1,079,345**	**1,239,392**	
36	0603285E	Advanced Aerospace Systems	3	38,093	153,700	246,008	U
45	0603739E	Advanced Electronics Technologies	3	213,379	189,564	180,408	U
48	0603760E	Command, Control and Communications Systems	3	129,182	115,148	110,101	U
49	0603762E	Sensor and Guidance Technology	3	138,508	192,086	224,860	U
50	0603763E	Marine Technology	3	25,290	32,487	33,360	U
51	0603764E	Land Warfare Technology	3	130,610	153,087	162,108	U
52	0603765E	Classified DARPA Programs	3	96,716	158,895	275,899	U
		Advanced Technology Development		**771,758**	**989,967**	**1,222,508**	
103	0605114E	BLACK LIGHT	6	4,240	5,000	5,098	U
113	0605502E	Small Business Innovative Research	6	42,363			U

PAGE D-2F

UNCLASSIFIED

UNCLASSIFIED

Defense Adv Research Projects Agcy
FY 2003 RDT&E PROGRAM

EXHIBIT R-1

APPROPRIATION: 0400D Research, Development, Test & Eval, Defwide

Date: FEB 2002

Line No	Program Element Number	Item	Act	FY 2001 (Thousands of Dollars)	FY 2002 (Thousands of Dollars)	FY 2003 (Thousands of Dollars)	S E C
121	0605898E	Management Headquarters (Research and Development)DARPA	6	32,944	38,102	43,572	U
		RDT&E Management Support		**79,847**	**42,102**	**48,572**	
124	0909999D	Financing for Cancelled Account Adjustments	6	1,008			U
		RDT&E Management Support		**1,008**			
		Total Defense Adv Research Projects Agcy		**1,976,682**	**2,252,717**	**2,685,110**	

Figure 6.2 DARPA's 2001-2003 Budget

A better example would be to look at the NASA budget. NASA maintains about $15B per year as its operating budget. The Shuttle takes up about a third of that budget, operating other centers takes up a good portion of it, unmanned missions, and upkeep of things like the Hubble Space Telescope and the International Space Station will impact the NASA budget also. Of course NASA employs a workforce much larger than the *Sixth Column* will but some of the efforts it undertakes are from the concept through flight.

Another very good example is the National Reconnaissance Office (NRO). The NRO has claimed in open sources to have a budget that averages about $6B per year. The NRO has developed technologies and flight systems that might prove similar to some of the technologies investigated by the *Sixth Column*. Again, like the DARPA systems, the NRO may reach order of magnitude type technology breakthroughs, whereas the *Sixth Column* would require much larger breakthroughs.

So, a budget for research and development that is somewhere between the DARPA and NASA budgets and maybe on the order of the NRO budget might prove to be satisfactory for the *Sixth Column*. We will assume for now that the budget line will be four and a half billion dollars per year (the average of the DARPA and NRO budgets). Table 6.1 shows a breakdown of the budget for the first few years of the *Sixth Column* once it has reached its fully operational form. Note that the initial cost is somewhat larger than subsequent years due to the construction and procurement of the infrastructure.

Table 6.1. Rough Order of Magnitude Budget for Sixth Column's first 3 years of fully operational status

	Year 1	Year 2	Year 3
Personnel	$120M	$120M	$120M
Upkeep of Infrastructure	$60M	60M	60M
Research Budget	$4,500M	$4,500M	$4,500M
Infrastructure Procurement	$300M	$0	$0
Total	$4,980	$4,500M	$4,500M

As mentioned previously however, the *Sixth Column* organization should start as a small group of less than a dozen or so very talented individuals. These individuals would develop the core philosophies, mission goals, general approaches, and standard operating procedures that the fully operational *Sixth Column* could implement. This core group might consist of the future DSC, Chiefs, and the DDSCs as well as other individuals suited for the project. Assuming that the "Core" team consists of a dozen people, then a budget of about $4.8M would be required for salaries and overhead per year. Offices, security, computers, and various other infrastructure requirements would suggest that another $5M or so would be required per year.

Also, a research budget of a few million to a few tens of million would improve the Core's likelihood for success. Therefore, a budget of less than $20M per year would probably be sufficient for the initial phase for the Core to begin building the *Sixth Column*. No more than five to ten years should be allowed for the building phase to be completed. The longer we drag our feet with this project the less prepared we will be if the ET invasion ever shows up.

6.5 Patriotic, Ethical, and Legal Responsibilities of the *Sixth Column* Members

Just as we believe that it is necessary to pay members of the *Sixth Column* at a rate higher than the average, we believe that the members of the *Sixth Column* have some ethical responsibilities to our society. These people are after all the potential defenders of mankind. We believe that it is imperative that these individuals know and obey the laws of our land. We also believe it is imperative that they know and obey the social ethical norms.

The members of the *Sixth Column* will have to sign an oath attesting to their allegiance and loyalty to the laws, ethics, and social norms of our country. They must affirm in the oath that they will protect the interests of their fellow man, and they will be required to reaffirm that oath every three years.

Obviously, they must protect the security of data and other important information such as designs with which they're entrusted. They must be absolutely committed to this. We must be able to tell them to make the best decision based on an honest evaluation of all the data available. All the members must be trustworthy and would be expected to be an honest person.

No one who has a felony record would be allowed to be a member of the *Sixth Column*. A member of the *Sixth Column* would be a positive role model within the community in which he or she lives. Each would be expected to do the right ethical and legal thing even when no one was watching.

Prospective members of the *Sixth Column* would be advised of some minimum expectations in terms of behavior at the beginning of their employment. At the orientation they would be given general expectations along with a list of specific behaviors, which would be prohibited. There would be zero tolerance for the following and other identified abhorrent behaviors:

- drug abuse
- alcoholism
- cheating on spouses
- sexual harassment
- abusive behavior to family or fellow workers

- cheating on taxes
- excessive speed in an automobile or other roadway vehicle
- negligence to one's duties and one's responsibilities
- failure to make court dictated or contractually required payments such as alimony and child support
- violation of software and hardware usage agreements.
- showing favoritism toward selected contractors

Certainly this list should be expanded immediately and over time as seen reasonable by our government. There would be a onetime amnesty at the beginning of service from non-felony offenses. Most definitely, this list is notional and could be expanded or reduced.

The punishment for violating one's oath should be finitely specified at the orientation. A harsh but specific punishment should be defined for offense of each prohibited behavior. It is important that the members know exactly what is expected of them and exactly what will happen should they fail to perform in accordance with their training on the oath.

There should be an opportunity for one time forgiveness of lesser offenses. Repeat violations of those offenses and first-time violations of the rest should be dealt with harshly. The punishment should include lifetime banishment from membership in or via contract with the *Sixth Column*. Breach of secrecy would be covered by ridicule of the individual's claims. "Aliens at Area 51? That guy never even worked there. What is Area 51? What a crackpot!"

Members of the *Sixth Column* would be expected to execute procurement competitions fairly and honestly. All evaluation criteria and evaluation methods would be clearly spelled out in the request for proposals. Evaluations would be based upon a numerical average of the evaluators. The contestant with the highest score would be awarded the contract for executing against that procurement.

There would be no room for judgment by any evaluator of or member of the evaluators' management team. The job of the Source Selection Authority (SSA) would be only to validate the scoring by the evaluation team. The SSA could not change the outcome of the scoring by the evaluation team. This approach would require change to the Federal Acquisition Regulations (FAR) or at least an exemption from the FAR as currently written. "Best Value" as currently defined by the FAR would be eliminated for *Sixth Column* procurements. This would also increase the fairness of procurements. It would also force Government employees to do a better job of thinking through what they are planning. Ideally, "Best Value" would be eliminated eventually

National Industry Security Program Operating Manual (NISPOM) and the Director of Central Intelligence Directives (DCIDs) will be used as the template for security of the *Sixth Column* as these are the national standards for classified programs security and methods of conduct. The *Sixth Column* will operate similar to programs, which have access to so-called Sensitive Compartmented Information as defined by the DCID:

> *...Sensitive Compartmented Information (SCI) Classified information concerning or derived from intelligence sources, methods, or analytical processes, that is required to be handled within formal access control systems established by the Director of Central Intelligence (DCI).*

However, the information will be deemed SCI for the *Sixth Column* through the Director of the *Sixth Column* (DSC). But, the SCI for *Sixth Column* information will not follow the requirements of the DCID whereas:

> *All DoD Components with responsibility for protecting SCI will establish and foster partnerships with elements of the DoD SCI community, the DCI, industry, and professional associations to gain insight to establishing common security practices and reciprocity wherever practicable.*

The reason being, of course, that the rest of the SCI community has no "need to know" about the *Sixth Column* activities. In fact, leaking SCI *Sixth Column* activities would be one of the extreme violations that would require legal banishment from the program.

The Ethics requirements mentioned above and within the NISPOM and the DCID only tell part of the story as to the caliber of people who are privileged enough and "super human" enough to work in the SCI community. As stated in both the NISPOM and the DCID, these individuals are subject to detailed legal and lifestyle background investigations. A person who typically cheats on a spouse would not qualify for SCI *Sixth Column* access.

How would such personal trait fallacies be discovered one might ask? As according to the rules of the SCI community, periodic and regular investigation, questioning, psychological evaluation, and polygraph testing is conducted. Therefore, only individuals who really do meet the requirements as governed by the SCI *Sixth Column* rules will actually be allowed within it. These types of individuals within the SCI community really and truly are the "good guys".

Individuals that persist in the science fiction genre like the leader of the "Majestic" organization in the television series *Dark Skies* or the "smoking man" in the series *The X-Files* or the ridiculous bad guys of *Alias* would never have been allowed into the *Sixth Column* program to begin with since they could not, would not, and do not meet the moral, legal, and ethical requirements of the *Sixth Column*, or any other SCI community for that matter. Of course, there are occasional "bad apples" but they are eventually caught and dealt with through legal means.

Typically, television and Hollywood efforts in general give the classified world a very bad reputation, which is unfounded, incorrect, and just plain stupid. One good television example of decent ethical folks in such a classified arena is the series *Stargate SG-1*, in which all of the individuals within the "Stargate Program" meet such "super human" moral requirements. It is funny that a series filmed in Canada is more "American" than most series filmed in America. Interestingly enough with the *Stargate SG-1* series is that the unethical folks are usually the elected and appointed officials like Senator/Vice President Kinsey. A good book example of a good secret organization is the organization in *The Puppet Masters* by R.A. Heinlein.

6.6 *Sixth Column* Research & Development Efforts – Possible 1st Funded Programs

Section 6.6 is designed to offer examples of the types of technologies that might prove useful in the event of an ET invasion provided that proper resources were expended on their development. This section is not meant to be an exhaustive list and description of the technologies. Instead it gives a few examples of the types of technologies that might prove useful and that should be developed. These are the types of technologies that the *Sixth Column* might be interested in developing.

Railguns

This section is not designed to be a "how to" on railguns. Instead it is assumed that the reader has some knowledge of the devices. We discuss railguns here because of their potential for creating very high energy projectiles.

Consider the simplified railgun shown in Figure 6.3. A high voltage power supply is used to charge a capacitor, which is then discharged through a sliding bar sitting on two conducting rails. There is a constant magnetic field B_0 in the (-**z**) direction. Due to the Lorentz force of the magnetic field acting on the current flow in the sliding bar, the bar is pushed in the (**x**) direction.

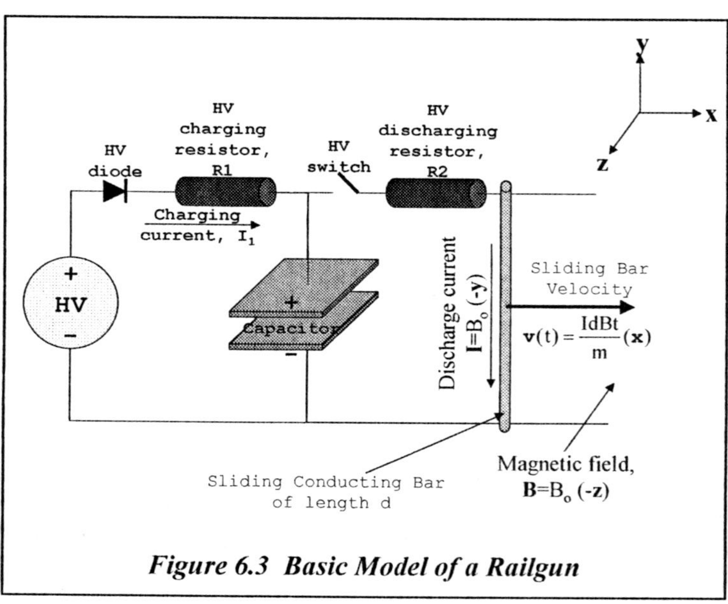

Figure 6.3 Basic Model of a Railgun

The force, F, on the bar is

$$\mathbf{F} = m\frac{dv}{dt} = \mathbf{I}\mathbf{d} \times \mathbf{B} = IdB_o(\mathbf{x}) \quad (6.1)$$

where m is the mass of the bar, I is the electric current in amperes, d is the length of the bar in meters, and v is the velocity of the bar. Integrating Equation 6.1 and solving for the velocity yields

$$\mathbf{v}(t) = \frac{IdB_o t}{m}(\mathbf{x}). \quad (6.2)$$

Realizing that the velocity is along one axis we can consider Equation 6.2 a scalar or

$$v(t) = \frac{IdB_o t}{m}. \quad (6.3)$$

Integrating Equation 6.3 once more gives the position as a function of time as

$$x(t) = \frac{IdB_o t^2}{2m}. \quad (6.4)$$

Assume that our railgun is to be space based and also of a reasonable and somewhat practical size. It is likely that future space navy ships could be as large as 100 m in length and maybe more. So we will assume that we could have a total railgun length, x =100 m. In other words, the total distance the sliding bar can be accelerated is 100 m. Also, a reasonable length d of the bar is 1 cm or 0.01 m. A sliding bar only 1 cm long might have a mass of about 1 g (0.001 Kg) depending on its material. The railgun can implement permanent rare-earth magnets (such as cobalt magnets) along its length which would apply a steady field of B_o~1 Tesla. So the total time it would take for the projectile to traverse the railgun's total length is then dependent only on the current. The current flow from the discharging capacitor is a decreasing exponential. If the value of the discharging resistor, R2, is chosen properly, the current flow can be fairly constant during the time the sliding bar is traversing the length of the railgun. The total time required for the sliding bar to traverse the railgun is

$$t = \sqrt{\frac{2xm}{IdB_o}}. \quad (6.5)$$

Figure 6.4 shows a graph of the time required for the sliding bar to travel the railgun's length versus the constant discharge current value. The graph considers current values from 1 KA to 1MA. The travel times are all much less than a second.

Substituting the time in Equation 6.5 into the velocity Equation 6.3 yields

$$v(t) = \frac{IdB_o t}{m} = \frac{IdB_o}{m}\sqrt{\frac{2xm}{IdB_o}}. \quad (6.6)$$

The kinetic energy of the sliding bar is

$$E = \frac{1}{2}mv^2 = \frac{1}{2}m\left(\frac{IdB_o}{m}\sqrt{\frac{2xm}{IdB_o}}\right)^2. \quad (6.7)$$

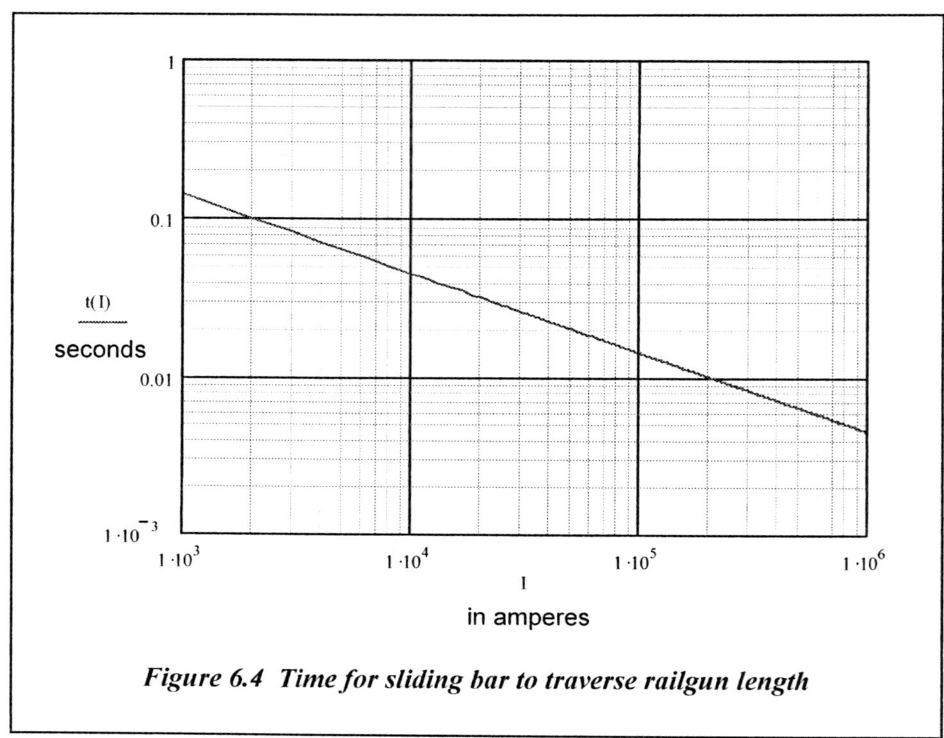

Figure 6.4 Time for sliding bar to traverse railgun length

Simplifying Equation 6.7 results in

$$E = xIdB_o . \tag{6.8}$$

Figure 6.5 shows a graph of the kinetic energy of the sliding bar after it is pushed the full 100 m (as shown in Equation 6.8) as a function of discharge current. Note that even with a 1 MA current source the energy of the projectile is only a megajoule. Recall that the ETs manipulate energy at much higher levels. So, are railguns useless?

Perhaps not. It is possible that larger magnetic fields could be achieved. If superconducting magnetic coils were used, the magnetic field could be an order of magnitude higher. This would only increase the projectile energy to 10 megajoules or 1×10^7 J. Substantially longer railguns would be required in order to achieve ET scale energies. And what about making the particles smaller or larger? What if the bullet could be large enough to carry a nuclear payload? Concept analysis needs to be done.

Although a system as described above is simple as far as the governing physics is concerned, there are significant engineering difficulties in constructing such a space-based railgun. Therefore, making one even larger will increase the engineering requirements.

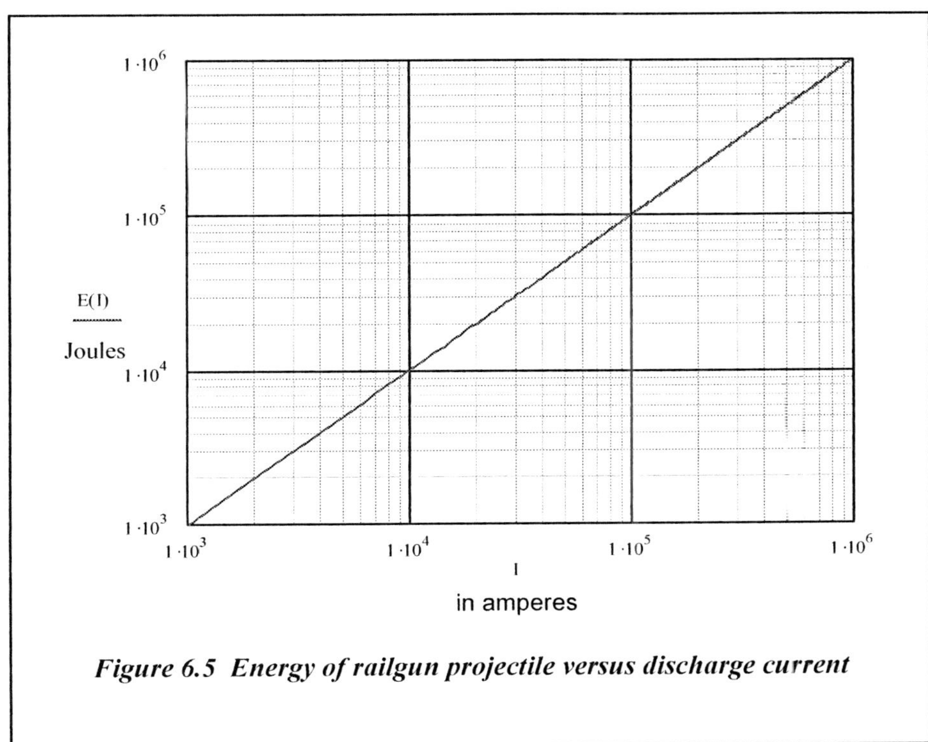

Figure 6.5 Energy of railgun projectile versus discharge current

It should also be noted here that these types of systems could have large projectile magazines. In other words, the railguns could fire many times like semiautomatic weapons. Although, one projectile can deliver only a megajoule or so, it is possible that the system could fire hundreds of times per minute, maybe more. This becomes an engineering problem. How long does it take to charge up the storage capacitor? Can an array of multiple capacitors be used as multiple discharge circuits in order to increase the fire rate? Could many railguns be placed on one spacecraft? These questions and many more need to be addressed.

Consider an array of railguns. Since the sliding bar (projectile) is only a centimeter on a side, the hardware supporting the rails is what drives the rail dimensions. Of course the power supply and capacitors might get large, but that could be the bulk of the spacecraft. So, if the rails take up as much as a 1 meter square surface (overestimating) then an array of 50 by 50 railguns would require a spacecraft 50 by 50 by 100 meters plus another container for the power supplies, capacitors, and control electronics. However, this suggests that a spacecraft the size of a couple of aircraft carriers could support a railgun array with 2500 railguns. If all of these railguns were fired at once and at maximum power, as much as 2.5×10^{10} J could be delivered to a target. Also, if the electronics were worked out in such a way that the array could be fired at a rate of ten times per second then another order of magnitude of energy would be gained.

A system like this would require large energy sources and a complex infrastructure, but it is not impossible. Although the energy level is still much smaller than the average α_2 ET, it is possible that an armada of these type devices good bring to bare orders of magnitude more energy. Or, future developments might enable much longer guns with higher electric currents and magnetic fields. Also, what happens to the attitude of the spacecraft when these high-energy particles are ejected from it? Many questions about this technology need to be answered. Research on these devices should continue.

Many other technologies must be considered and the list is far too great to discuss each of them in one chapter. However the following are examples of other technologies that should be investigated by a *Sixth Column* type organization.

- Space Launch Earth-to-orbit technologies
- In space propulsion concepts
- National Missile Defense
- Tactical, ground, air, and space based directed energy weapons
- New types of armor (Powered Armor and Mecha as discussed in Chapter 2)
- EMI weapons
- Micro, nano, and pico satellites
- Nanotechnology
- Advanced Software Weapons
- Directed Solar Energy
- New Energy Sources
- Defensive shields
- Ultra large arrays for optics and RF
- Lunar Catapults
- Space Based Missile Platforms
- Space, Lunar, Mars, etc. Bases
- Space Navy
- Civil Defense Shelters
- Others yet discovered.

The Most Serious Deficiency in Man's Planetary Defense Capabilities

What is the most serious deficiency in Man's planetary defense capabilities? The answer is obvious: space propulsion. At present, no one on Earth can conduct even the most limited military operations in space, even inside the Earth-Moon system. The reason is that no remotely adequate propulsion systems exist. At least as far as the public knows, no one is making any serious effort to rectify this situation.

If the Solar System were invaded, Man would be in the unenviable position of being forced to defend his home planet on that planet. History provides us with many examples of nations that were forced, for one reason or another, to fight defensive wars entirely within their own borders. Sometimes they were successful, sometimes not, but they invariably suffered heavy civilian casualties and heavy economic losses.

On the other hand, we have the example of England. No enemy has landed on England's shores in more than nine hundred years, largely because of her naval superiority. Of course, naval superiority will not protect you from invasion if you have enemies on your borders--but a planet is, like Britain, an island.

It is a rule of thumb that a space propulsion system for a given mission should deliver an exhaust velocity roughly equal to the total velocity required for the mission. By this criterion, chemical-propellant rockets are not adequate even for orbital missions; hydrogen-oxygen engines have exhaust velocities of about 4.5 km/s, but it takes about 9 km/s to get into orbit, including drag and gravity losses. This is why we use multi-stage rockets.

With chemical-propellant rockets, we can carry small payloads into orbit. That is the limit of our capabilities. We could put *very* small payloads such as bombs onto low-energy escape trajectories, but such weapons would surely be completely ineffective against an alien space fleet, or against an alien base on the Moon.

Solid-core nuclear rockets, which we could build right now, could deliver exhaust velocities of at least 10 km/s at high thrust/mass ratios. They would be quite adequate for orbital missions. Bearing in mind that it really doesn't make sense to use the same ship for surface-to-orbit and orbit-to-orbit missions, they would even be adequate for missions inside the Earth-Moon system.

We could build real spaceships, if only short-range ones, right now, and if we were doing so, our military position would be far, far better.

For the exploration of the inner Solar System, solid-core nuclear rockets would be fine, but for operational missions, we need something better. We are reasonably sure that we can build gaseous-core nuclear rockets delivering exhaust velocities of 30 km/s at high thrust/mass ratios. Missions in the inner System would be no longer than naval missions in sailing-ship days.

At higher exhaust velocities, gaseous-core nuclear rockets are in competition with ion drives and plasma jets. Which you favor depends upon whether you think a gaseous-core nuclear rocket with an exhaust velocity of 200 km/s would have a higher thrust/mass ratio than an ion drive or plasma jet with the same exhaust velocity.

The capabilities of thermonuclear propulsion systems are speculative, but something resembling Robert A. Heinlein's torch ship is by no means out of the question.

In our opinion, weapons development deserves much lower priority than propulsion system development. For all we know, weapons that would curl science fiction writer Doc Smith's hair may be possible (Doc Smith's stories often included weapons that could destroy entire star systems and more). And, for all we know, the weapons we have today might be as ineffective against invaders as spears against a battleship. But, while we can clearly foresee space propulsion systems that would compare to those now in use as the F-15's engines compare to the engine in Henry Ford's Model T, we cannot clearly foresee weapons vastly better than those we already have without major research and development efforts. It does not make sense to expend a great deal of effort on advanced weapons with no means of implementing them. In other words, the mightiest weapons are of no value if we cannot get them to where they are needed.

The further development of sensors and computers seems even less urgent. During the debate over missile defense in the 1980s, it was sometimes said that an effective defense would require impossible sensor and computer capabilities. But, to the extent that this was true, it was a consequence of the arbitrary and unrealistic constraints that were imposed upon the system – for example, that interceptor missiles could not have nuclear warheads. There is no reason to think that our planetary defense capabilities will be in any way limited by sensor and computer technology.

Breakthrough Physics Propulsion

In order to defend against an enemy from space, it would be beneficial to be able to reach space. Presently there are only a few ways to do that. The U.S. has several expendable rockets as well as the Space Shuttle. The Space Shuttle can lift thirty thousand kilograms (sixty thousand pounds) to orbits of only a few hundred miles above the Earth. It is not useful for higher orbits but is a workhorse for carrying large amounts of mass to the lower ones.

The expendable launch vehicles (ELVs) are our only present hope for reaching higher orbits. The Delta IV Heavy, for example, can push payloads of more than thirteen thousand kilograms to a highly eccentric orbit known as geo-transfer orbit (GTO). This is an elliptical orbit with the periapsis at low Earth orbit (LEO) of about a thousand kilometers (six hundred twenty miles) and an apoapsis at geosynchronous Earth orbit (GEO) of about thirty six thousand kilometers (about twenty three thousand miles) above the Earth.

The payload shroud for the Delta IV is about five meters in diameter and twenty meters tall. It is possible that it a small manned vehicle could be mounted within the Delta IV Heavy shroud and launched to GTO. Motors on the vehicle could be used for spaceflight in and around GEO and back to Earth. In fact, this is the approach that NASA has chosen as a near term Space Shuttle replacement. An expendable rocket much like the Delta IV Heavy (if not actually the Delta IV Heavy) is to be used for main propulsion for the Orbital Space Plane (OSP), which is a lifeboat and ISS crew return/transfer vehicle that will be mounted atop the ELV. The OSP will be too heavy to reach orbits much higher than the ISS and might prove fairly useless for planetary defense. However, the concept is viable. The new approach is to consider using the Shuttle solid rocket boosters (SRBs) with an upper stage that uses the Space Shuttle Main Engine (SSME). It appears that this concept will be the one that moves forward.

If the ELVs are used to launch small one, two, and three man spacefighters/bombers and even unmanned missile batteries to higher Earth orbits, there may be some benefit to the defense of the planet. Sixth Column efforts should include a development effort of this technology.

If the aliens were to base themselves further away from Earth than GEO, then there is no quick technological approach of delivering a manned vehicle to the location. Man has not traveled higher than a thousand kilometers above the Earth since the last Apollo mission and that technology is long lost.

New In-Space transportation development efforts need to be a major portion of the Sixth Column's mission. The current efforts of NASA Marshall Space Flight Center's In-Space Propulsion Program should be leveraged although most of the technologies NASA is pursuing deliver high specific impulse and low thrust. Examples of the NASA In-Space Propulsion Program development efforts are:

- solar sails,
- solar thermal,
- ion propulsion,
- tethers,
- and advanced chemical propulsion.

For quick delivery and combat capabilities high thrust would be a requirement. Therefore, a new development effort must be undertaken. The most likely candidate to date would be nuclear thermal propulsion (NTR), whereas a nuclear fission reactor is used to superheat a propellant through a converging diverging rocket nozzle system.

NTR could deliver orders of magnitude greater thrust than the Space Shuttle engines and many many orders of magnitude over currently funded In-Space concepts. However, NASA has, for some reason, an aversion to NTR systems. The reason is purely political and not technological. The *Sixth Column* should not need to deal with such worries and should be able to develop the proper technology for the global defense task requirement. NTR is the near term development effort that has most likely payoff. Until new technologies are developed chemical systems will have to suffice.

Another part of the *Sixth Column* mission would be develop next generation and futuristic advanced propulsion concepts such as are allegedly pursued by the NASA Breakthrough Physics Propulsion program or BPP. These technologies would be: warp drives, stargates, antigravity, quantum foam transporters, and any other science fiction sounding concept you can think of. However, some of these are viable and could possibly be developed to useful stages within a few decades if the proper funding is released. On the other hand, we can't expect NASA BPP will do this. The BPP budget has been roughly five hundred thousand dollars per year since the program was created in the late nineties. Five hundred thousand dollars will fund roughly two individuals per year in modern corporate America. Two guys, no matter how smart they are, probably will not build the world's first faster than light engine or antigravity machine. And more recently, the NASA BPP budget has been practically zeroed.

From 2000 to 2002 NASA spent about five billion dollars on the so-called Space Launch Initiative or SLI. The SLI program was to develop the next Earth to Orbit spacecraft that would replace the aging Shuttles. Dr. Taylor was present at the NASA Advisory Council Meeting on SLI and observed the council to say that, "We've spent this much money and all we've got is a bunch of pretty vugraphs and animations!" This is mostly true for the SLI development effort.

Then in 2003 NASA was getting ready to actually attempt to direct some of those vugraphs toward real hardware development, but they changed their minds. Instead, they decided to start all over and develop the Orbital Space Plane (OSP), which is back at the pretty vugraph phase. This is nearly what happened with the X-33 program where billions were spent for vugraphs and some faulty and heavy composite fuel tanks. Then it was decided to start over and develop SLI. Then SLI went nowhere and OSP has emerged. Will anything ever get built for these billions and billions of dollars? With the advent of the Columbia disaster and President G.W. Bush's new space initiatives, and a new NASA administrator in 2005, we are likely at the vugraph reset stage.

Had NASA spent half of the X-33/SLI/OSP budgets on BPP they would at least have created just as pretty vugraphs of warp drive engines rather than just Shuttle replacements that have little chance of ever flying. For a real successful BPP effort, an organization like the Sixth Column must set up a BPP development program with specific goals, timelines, and management controls. The Manhattan Project would be an excellent template from which to work. With resources for the BPP on the order of SLI, the nations top General Relativists, Quantum Mechanics, and Wormhole Construction Workers could be forced together into a closely ordered and controlled and funded environment. In 1939 nobody believed in an atomic bomb. After to the Manhattan Project the Japanese believed in them in a major way.

The bottom line is that for a global defense we cannot place our hopes in NASA to develop space propulsion technologies. NASA as a civil agency is not supposed to develop uniquely military or weapons technology. The *Sixth Column* must take charge of the space propulsion development efforts and be given a large untouchable budget to fuel those efforts.

Intelligence Collection – SIGINT, IMINT, MASINT

We need to collect intelligence to the extent possible about ET. Ideally, we would do this without telling ET anything about us. Individuals frequently collect intelligence about others. One may do this in person or by hiring a detective. Countries collect intelligence against others routinely. We have agencies such as the Central Intelligence Agency (CIA), The National Security Agency (NSA) and the National Reconnaissance Organization (NRO) to collect intelligence for us as a nation.

It is a good thing that we have them as their activities have saved countless American lives. These organizations often collect data remotely. We will, at least for the near-term future, need to collect intelligence against ET remotely. This mission should fall under the *Sixth Column*.

A common technical method used to collect intelligence against others is SIGINT. SIGINT is an acronym for Signals Intelligent. SIGINT is commonly collected in the RF spectrum. Optical techniques can and are also be used to collect SIGINT. SIGINT is routinely collected through the use of remote radios. There are several areas of technical collection within the SIGINT area. The two which are most practical for application against ET are ELINT and COMMINT.

ELINT detects the presence of electronic emissions and attempts to geolocate them. Collection and geolocation will prove challenges over the ranges that collection systems will have to operate. COMMINT, which attempts to copy voice and data transmissions, will also face a significant challenge because of range. The collector must be able to either see a main beam or be in a near in side lobe to collect against it.

COMMINT analysts will face another huge hurdle in the area of linguistics. They must be able to interpret the signals collected. We should anticipate that ELINT will be significantly easier at least until we have made great strides in linguistics.

While SETI rarely, if ever, discusses the use of SIGINT, it uses RF based SIGINT extensively in its search for signals from ET. SETI is relying on ET to use SIGINT techniques to receive its transmissions.

Most American households use SIGINT techniques everyday. Every time a radio or television is turned on SIGINT techniques are put to use for the benefit of the user. Most people refer to these radio, television and telephones as communications devices. The term SIGINT is typically reserved for applications in which signals are collected without the consent of or knowledge by the transmitting body. Communications are typically performed with the consent and cooperation of the transmitter. There is a reasonably high probability that if we collect signal from ET that it will be SIGINT as it will be noncooperative or accidentally cooperative.

We should consider it cooperative only if it clearly is a response to one of our transmissions. We still recommend that we stop deliberately transmitting. But we must remember that even a cooperative response may come from a hostile ET. Based on the little we know today, there is an approximately equal probability of receiving a response from a hostile ET as from a benevolent one.

We are somewhat limited by our technology for the selection of SIGINT systems. Figure 6.6 depicts the typical components of a SIGINT collection system; it is a radio. The components to build a SIGINT collector can be purchased quite readily from amateur radio electronics stores or online. However, they need to be integrated in a fashion that tailors them for the specific application.

Each of the elements in the diagram must be tailored for the band of frequencies and the type of signal that you intend to monitor. The antenna system can range from a simple whip or long wire for shortwave signals (3-30 MHz) to elaborate arrays or large aperture dishes for microwave ranges (3-30 GHz). Many commercially available radios can meet the demands of a SIGINT system for frequency agility, multiple demodulation modes, trunking and computer control. Many are available as a "brick" without knobs and dials (Ten-Tec, ICOM, Yaesu) and are specifically designed for use with a PC.

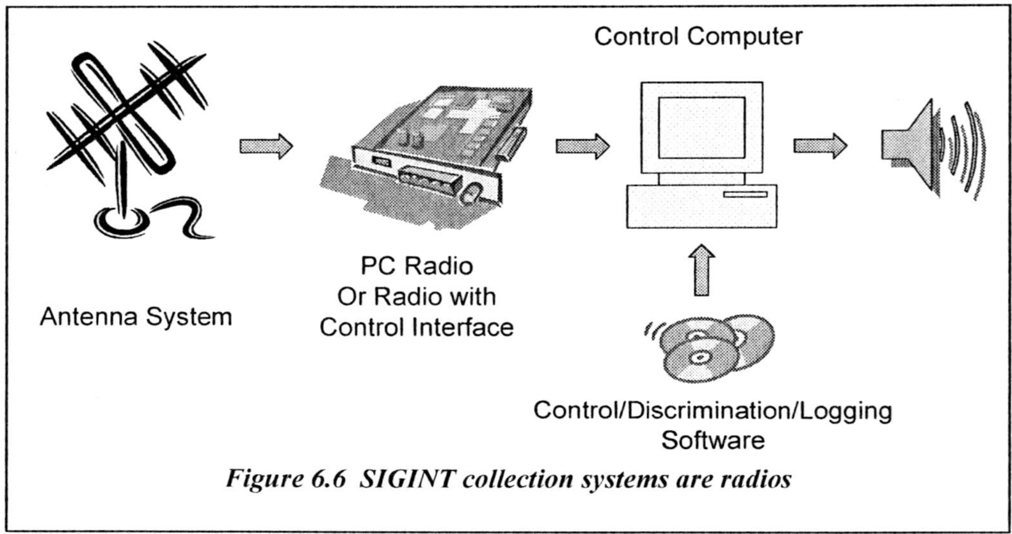

Figure 6.6 SIGINT collection systems are radios

Others, such as the WiNRADiO®, install directly into an open slot on your PC motherboard and become an integral part of the PC. The last item is software. Software for radio control, data logging, spectrum analysis, detection, discrimination and various algorithms for noise cancellation are available commercially and form the user interface for controlling the SIGINT station. Since the SIGINT station described is PC based, it can easily be networked and become a node on a much larger grid of monitoring stations that cover different bands or targets.

We use only a small portion of the electromagnetic spectrum for collecting SIGINT. The electromagnetic spectrum is a continuum of all electromagnetic waves arranged according to frequency and wavelength. The sun, earth, and other bodies radiate electromagnetic energy of varying wavelengths.

When you listen to the radio, watch TV, or cook dinner in a microwave oven, you are using electromagnetic waves. Radio waves, television waves, and microwaves are all types of electromagnetic waves. They only differ from each other in wavelength. Waves in the electromagnetic spectrum vary in size. They range from very long radio waves the size of tall buildings to extremely short gamma-rays which are smaller than the size of the nucleus of an atom.

While electromagnetic waves can be described by their wavelength, they can also be described by their energy and frequency. The electromagnetic spectrum includes, from longest wavelength to shortest: radio waves, microwaves, infrared, optical, ultraviolet, X-rays, and gamma-rays.

Most of the electromagnetic spectrum is invisible, and is composed of frequencies that traverse its entire breadth. Exhibiting the highest frequencies are gamma rays, x-rays and ultraviolet light. Infrared radiation, microwaves, and radio waves occupy the lower frequencies of the spectrum. Visible light falls occupies a rather narrow range in between.

Radio waves, such as those that FM radio stations broadcast, are merely electromagnetic waves with a somewhat lower frequency and longer wavelength than visible light. The millimeter radio waves, or microwaves, possess a slightly higher frequency and shorter wavelength than do FM radio waves.

For convenience in talking about the electromagnetic spectrum, we classify photons of different energies into different spectral regions. The photons in each of these regions exhibit the same electromagnetic nature. However, because of their very different energies they interact with matter quite differently. A representation of the divisions within the electromagnetic spectrum is shown in figure 6.7.

We are somewhat limited by our technology for the selection of SIGINT systems. The components to build a SIGINT collector can be purchased quite readily from electronics stores or online. However, they need to integrated in fashion that tailors them for the specific application.

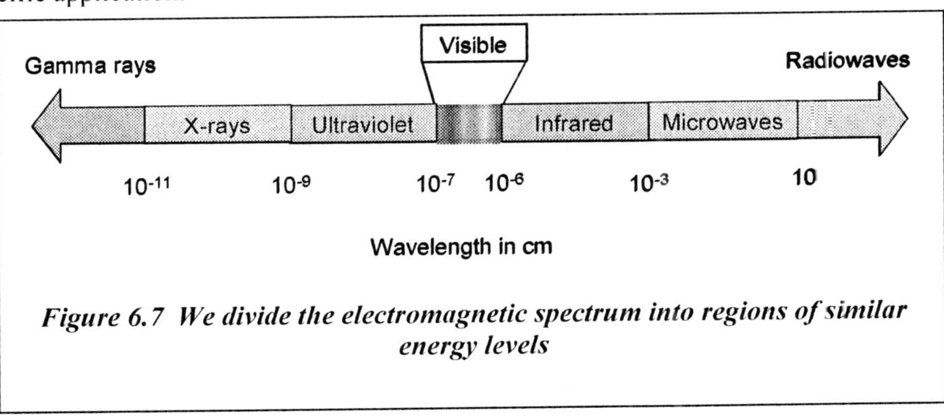

Figure 6.7 We divide the electromagnetic spectrum into regions of similar energy levels

Another technology that is emerging is the software defined radio. Computer hardware technology has now increased to the point that software can execute the mathematics defining the functions of a radio receiver faster than those functions occur in a typical radio circuit. What that means is that you can implement the same machine (i.e. a radio) using software. A software defined receiver is a radio whose waveforms are defined in software. Waveforms are generated as sampled digital signals by being converted from analog to digital via a wideband Analog to Digital Converter (ADC). The receiver algorithm then filters, discriminates and demodulates the waveform using software on the computer. A group of enthusiasts operates a web site at http://www.gnu.org/software/gnuradio/index.html and makes the source code available for download. This form of radio is very attractive for SIGINT because it allows for full and complete control over the algorithms used for demodulation and detection. The drawback is that performance is bandwidth limited in that the speed of the processor and ADC bound the monitoring frequency.

There are several obstacles to intercepting ET's emissions. Among those obstacles are:

- Range over which we must collect
- Frequency bands available to us
- Geolocation issues
- Language barriers

The biggest known enemy of SIGINT collection against ET is range, distance. The SIGINT systems we employ today have fairly small apertures as a general rule. Successful collection is based on having a suitable power X aperture solution to overcome range loss and background emissions. Frequency of the system also influences the need for aperture. The aperture size is driven by the ratio of the square of the frequencies; as frequency increases, the required aperture become smaller. There has to be high enough power and large enough aperture between the two ends of the link to make it close. The reason we can have a small radio antenna in our homes and on our cars is that the transmitting end is very high power. However, as we drive we still get outside the range of radio stations.

We will consider of the systems with the largest antennas available today. The NASA Tracking and Data Relay System have two 4.8 meter apertures. The ACeS satellites have 15 meter apertures. All other public systems except those of SETI are even smaller. Let us consider the SETI radios at Arecibo, in Puerto Rico and operated by Cornell University, and Effelsburg, in Germany.

Arecibo has the largest antenna. It is 305 meters in diameter. However, it is constrained to operate at <10 GHz, .327 to 9 Ghz, by its surface roughness of 2.2mm root mean square (RMS). As frequency increases, the gain of an antenna increases. Unfortunately, surface roughness degrades gain faster than frequency increases it. Effelsburg is smaller at 100 meters in diameter. However, its top operating frequency is significantly higher at 86 Ghz. It has higher gain than does Arecibo at high frequencies. Even with the gain of apertures of this magnitude, we have only a small fraction of potential signals from within the Universe available to us. Figure 6.8 shows various SIGNIT type systems and the spectrum of those sensors.

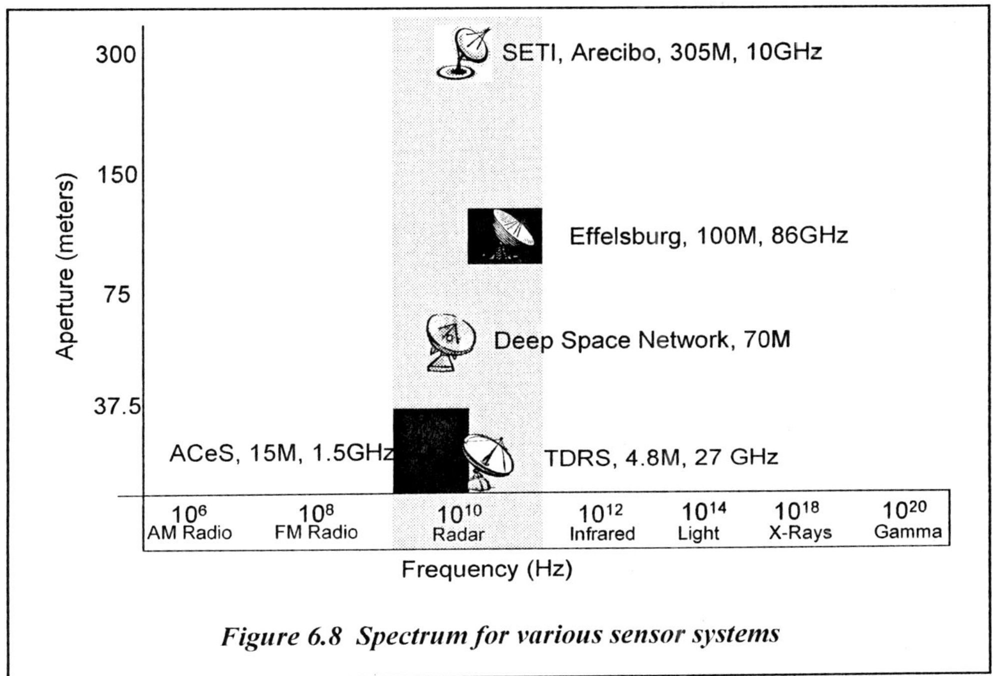

Figure 6.8 Spectrum for various sensor systems

The SETI community realizes the need for greater sensitivity to see further into the universe and to collect more photons. There are desires to build a system with 100 times the sensitivity of current systems. The desire is driven by the fact that background emissions from our galaxy make up a sizeable fraction of the total noise. This ambitious facility concept is called the Square Meter Array Radio Telescope. The technology concept selection is currently scheduled for 2005 with probably an operational system in ~2010 assuming technical agreements can be reached. Oh, and if $1B Australian can be raised. It is planned to operate in a frequency range of .2 to 12 Ghz.

You can see that if we used existing systems for SIGINT collection against ET, we would be taking advantage of only a small part of the electromagnetic spectrum as we know it. We even use a small part of the RF spectrum. All of the systems discussed except Effelsburg are limited to operating over frequencies below 16 Ghz. We are limited in frequency by:

- International Telecommunication Union (ITU)/World Radio Conference (WRC) rules and agreements
- Federal Communications Commission regulations.
- Antenna Technology status
- Component development and availability
- Component Manufacturing issues and difficulties

- Atmospheric propagation constraints.

It is statistically sound logic that at least some ETs will not operate in the same portions of the electromagnetic spectrum that we commonly use. This is true because they may not suffer from the same constraints. They are quite possibly without an ITU. They have developed their technology differently than we have; after all, our technology developments have been driven by the work of a few brilliant men and then confined to "normal science" for extension.

The ITU is viewed as a necessary body here on Earth. It serves to allow operators in all countries to be able to communicate within their allotted spectrum. Without the ITU, we Earthlings would be in a communications survival of the fittest. He could obtain the highest aperture X power product and or cancel out his competitors and would rule the spectrum.

Here is a case where the argument for free market control would aid technology development. Without the ITU, entities wishing to communicate would be forced to develop new and improved technologies. The ITU, while it seems helpful, is really an inhibitor to progress. The ITU is a tool of convenience driven by greed. In reality it was put in place to protect the interest of a, not the, few. We should abolish the ITU and let greed drive the vendors and Governments in the marketplace to develop technology or perish.

Antenna technology limits aperture size and efficiency. Increased aperture is the approach to collecting SIGINT against ET. It allows for ET having small aperture and/or low transmission power. Unfortunately, surface roughness is an issue in either case. As aperture grows it becomes increasing difficult to maintain, let alone improve, surface roughness. As frequency increases, surface roughness must be decreased to improve gain. There are practical limits to the minimum surface roughness achievable which current technology. Improving surface roughness is also limited by surface measurement techniques. We must have advances in both manufacturing and measurement techniques to obtain higher gain.

An alternative that offers at least some short-term advancement is the use of arrays. Arrays allow for manufacturing of smaller antennas where surface roughness is easier to control and minimize. Smaller antennas are easier to build, until they get so small that their manufacture becomes the equivalent of watch making, than larger ones. The individual antennas in an array can be operated independently or they can be operated as a single antenna. The alignment and surface roughness problems are now shifted to alignment and composite surface of the antennas in the array. There is also a need for digital signal processing and array control. We need additional work similar to that done by scientists such as Dr. Patrick Martin and Dr. John Shipley in antenna array technology in order to improve on it.

RF component technology that is needed to improve on our ability to collect SIGINT against ET is currently funded primarily by the commercial community. The few developments that occur are within those portions of the spectrum allotted by the ITU. There is little potential profit to be made by working in areas outside the current commercial arena. Components for high frequency (above about 40 Ghz) applications are typically built as custom items today because there is little demand.

Another factor limiting development for high frequency applications is that components become smaller as a ratio of the squares of the frequencies being considered. The requisite tolerances become tighter on higher frequency components to make them work properly. Tighter tolerances restrict the number of manufacturing techniques available for building components. A significant factor in the use of higher frequency systems is that of heat rejection. As power is applied to RF components they build up heat. Heat on small components may cause them to distort so that they fall outside the tolerances for proper operation. Smaller components have less thermal mass to absorb heat and less surface area to dissipate it by radiation. Small components therefore often require the use of active cooling. This may be done by using water or other liquids running through pipes along the surface. Phase change materials within or along the surface of components is another option as is cryrocooling. One control heating by limiting the time in operation; however, who knows when ET will speak or be available to be collected against.

It is difficult to fathom an environment in which ET does not face the same challenges we do. However, we should not rule it out. ET may have already solved the problems which limit our ability to collect SIGINT against him. He may have found techniques to reduce the power aperture requirements to allow small, low power radios to collect SIGINT against us.

We are not restricted to the use of SETI radios for SIGINT collection tools. The US intelligence community routinely builds and operates SIGINT systems. The National Reconnaissance Office shows a Directorate for Acquisition and Operation of SIGINT Systems on the organization chart shown on their website, www.nro.gov. While no details are given about the satellites the organization develops and operates, it is reasonable to assume they are built for SIGINT collection against sources on Earth. Their operating ranges are expected to be less than 35000 kilometers. That would be consistent with their stated mission as "FREEDOM'S SENTINEL IN SPACE."

Such collection systems would by design be smallest, most economical and easiest to launch if placed in orbits near the Earth. The launch vehicles owned by the US and other nations are restricted to on the order of 10,000 to 15,000 kilograms to Geosynchronous orbits (GEO). The throw weights of those systems diminish rapidly as orbital attitude increases. The power apertures of these systems are undoubtedly smaller than those of the largest SETI radios. The systems built by other members of the Intelligence Community are going to be even more constrained for use against ET as they are designed for far shorter ranges of operation.

A satellite system collecting SIGINT against ET offers a number of advantages. If we did launch a SIGINT satellite designed to collect against ET to Geosynchronous orbit, we could improve its sensitivity significantly over an Earth based system. The gain loss suffered range and atmosphere from space is substantial. We could recover that as sensitivity from space. Obviously, geosynchronous or higher orbit offers the greatest advantage. Unfortunately, as previously stated it offers the greatest launch obstacles. We would open up some collection frequencies which are unavailable to us from Earth because of atmospheric effects on radio wave propagation.

Even if we improve the power aperture of current collection systems, we will still have limited sensitivity and radio frequency spectrum available to us to collect SIGINT against ET. We need new technologies to expand our search for ET. Our SIGINT collection systems of today are much too limited in capability to find ET. Rain models are used to estimate the amount of degradation (or fading) of the signal when passing through rain. The degradation is primarily due to absorption by the water molecules and is a function of frequency and elevation angle. Generally speaking, the rain loss will increase with increasing frequency. The loss will also increase with decreasing ground elevation angle due to a greater path distance through the portion of the atmosphere where rain occurs. The rain will also cause an increase in the antenna noise temperature.

The rain models used in the Communications and Radar Modules are global annual statistical models. The annual rainfall rate and probability of the rate for a particular location are determined from historical measurements. In general, the world is divided up into different rain regions, each with its associated rainfall rates and probabilities. If both the transmitter and receiver are located above the rain height threshold, the rain loss is zero. Rain loss is computed for objects on the ground and for aircraft below the specified rain height.

We will need to develop new technologies for satellites to achieve SIGINT collection against ET from space. It is possible that we will find developing a space based system easier than developing an Earth bound system. It would be free from gravitational limitations. The real estate acquisition problem would be mitigated. By going to slightly higher orbit than GEO, we could minimize the satellite to satellite communications system interference which restricts satellites to two degree spacing from GEO; we would only have to solve the downlink interference problem.

Even if we place our SIGINT collection systems in geosynchronous or higher orbits, we are eliminating only a small fraction of the distance to extra-planetary celestial bodies. The path loss resulting from distance decreases only by a small fraction. However, we eliminate losses from atmospheric effects such as rain and turbulence which can be large contributors to the loss budget. We also open up RF frequencies which are denied by atmospheric effects for Earth bound SIGINT systems. We can significantly increase our chances of collecting against ET.

The fundamental aim of a radio link is to deliver sufficient signal power to the receiver at the far end of the link to achieve some performance objective. The benchmark by which we measure the loss in an RF link is the loss that would be expected in free space or the loss that would occur in a region which is free of all substances that might absorb or reflect radio energy. This is the ideal case which we hope to achieve in our real world SIGINT collection link. It is possible to have path loss which is less than that of "free space" but we should typically expect to experience greater path loss.

We will offer several concepts for building very large aperture SIGINT satellites. A satellite could be built in a GEO orbit using multiple launches to construct a huge SIGINT radio to collect against ET. An aperture built in this fashion could easily be an array. In fact, it would probably be a preferable solution. It would allow the largest aperture with the least volume and weight. Another concept would use a spacecraft, which would be distributed over and integrated into the structure of a large antenna. This is important because support systems and the satellite buss take up such a large fraction of the launch volume and weight.

The buss in today's satellites severely limits the amount of payload put on orbit by limiting the payload weight and volume fraction. In either of these approaches we would probably find it advantageous to have concentric frequency rings within the surface. Those frequency rings would start with the highest frequency closest to the center and frequency decreasing across the surface to the outer edge. This will allow achieving the most accurate surface in the area where it is most easily done; that being closest to the center of the structure.

The frequency of each subsequent ring would be reduced thereby requiring less surface accuracy. A third approach might be to use a swarm of satellites. Each of those satellites would have an aperture that would be reasonably small. The apertures on each of those satellites would be phased in such a matter as to make a large array. It would be necessary to have communications between the various satellites in the swarm.

One of the satellites would act as the communications center and would be responsible for collecting all ET data and relaying it to Earth. Communications between the satellites would probably the best accomplished at 60 GHz or using lasers to avoid interference with the communications of other satellites. The frequency plan of the downlink would be most challenging in order to avoid interference with the downlinks of other satellites.

IMINT – Imaging Extra-solar Planets

It is possible that planets around other stars will be imaged in the near future. Using modern photometry and Doppler techniques many extra-solar planets have been found in the last couple of decades. Technology is emerging that will even enable the enthusiastic amateur and backyard astronomer to enter the search for extra-solar planets. Once these planets are found, the task of imaging them and trying to understand their makeup and if they harbor alien life will arise. The technology to image extra-solar planets is not yet within the engineering ability of modern systems. Although the physics is well understood, there are many large hurdles in engineering that must be overcome. It is these types of technical hurdles that the *Sixth Column* should be interested in clearing. The following is a brief description of the state of the art in planet imaging and actual requirements for future imaging that might be needed for the *Sixth Column* to gain intelligence on the whereabouts of possible ally and hostile alien cultures.

The Hubble Space Telescope (HST) can image Pluto when it is at its perigee (30AU), which it is passing through until about 2015. Observing the planet while Earth is closest to it at 29 AU yields an image as shown in Figure 6.9. Figure 6.9 was taken by HST by Principle Investigators Stern, Buie, and Trafton (Stern & Mitton 1999). The image in the upper left shows the image taken by HST. The larger image in the center of the frames are computer enhanced. Note that the pictures give some idea as to the surface bolometric albedos. The two images are at separate times during Pluto's rotation.

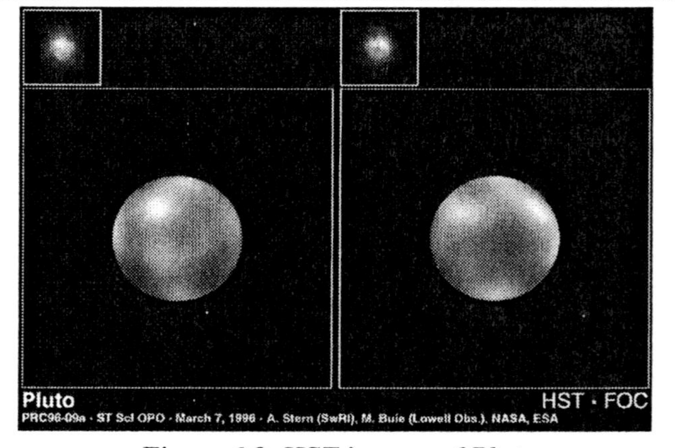

Figure 6.9 HST images of Pluto

Consider Pluto's distance and the resolution of HST in the following equation,

$$D = 2z \tan\left(\frac{\theta}{2}\right) \quad (6.1)$$

D is the minimum resolvable distance, z is 29 AUs, θ is .1 arcsec. Solving 6.1 yields a minimum resolvable distance D = 2108 km. Pluto's diameter is about 2300 km (Stern & Mitton 1999). Actually, realizing that the diameter of the HST primary is about d = 2.4 meters and using the Rayleigh criterion (Goodman 1998)

$$D = 1.22\frac{\lambda z}{d} , \quad (6.2)$$

and choosing 550 nm as the wavelength yields a resolution of about 1216 km. Hence, several pixels per image of Pluto can be obtained. After computer enhancement various important scientific measurements such as albedo change with time of day can be made. However, the resolution at the Pluto distance is still only a couple of thousand kilometers.

Now consider an Earth sized planet at the distance of the nearest star Alpha Centauri, which is about 4.5 light years away. Using equation 6.2 and solving for the aperture diameter, *d(z)*, as a function of the distance to the star, wavelength, and setting D equal to the diameter of the Earth, yields the results shown in Figure 6.10. Figure 6.10 shows that a single aperture telescope with a diameter of between 1000 and 2000 meters would be required to resolve an earthlike extra-solar planet in the Alpha Centauri system! This is orders of magnitude greater than HST and even the Next Generation Space Telescope (NGST). The engineering ability to develop such a Gossamer structure does not yet exist on this planet. Some engineering studies have been conducted for such structures for very large telescope applications and beam steering applications with laser sails. However, most of the studies conclude that the engineering state-of-the-art is not quite up to the task.

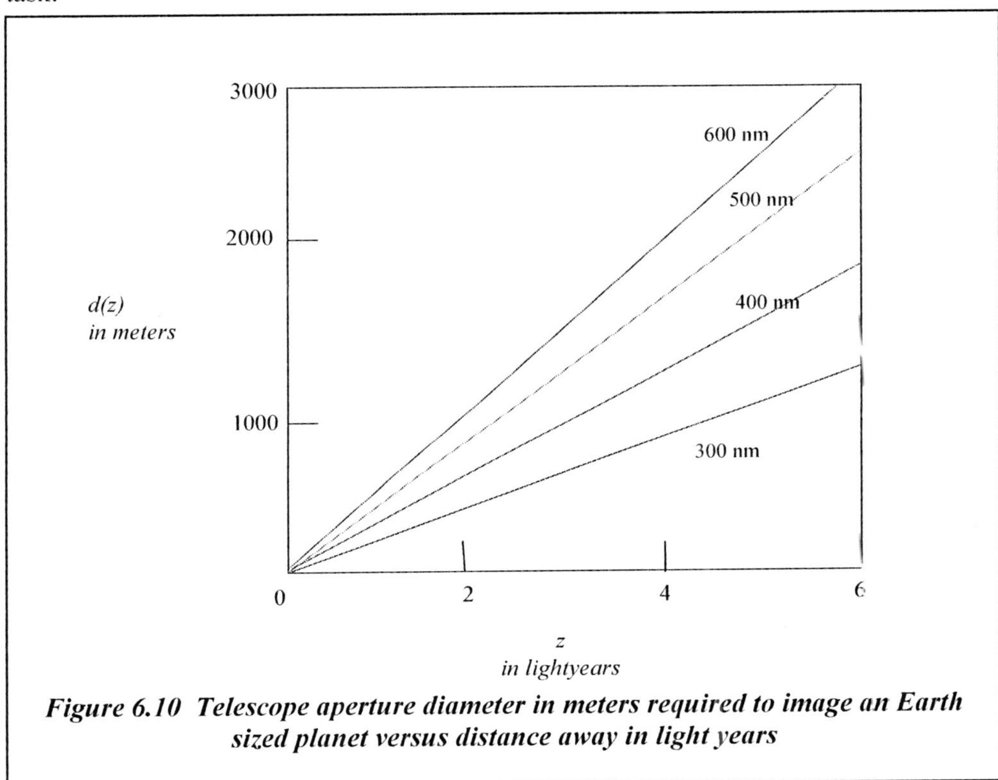

Figure 6.10 Telescope aperture diameter in meters required to image an Earth sized planet versus distance away in light years

So how could extra-solar planet imaging be accomplished in the near future without such Gossamer structures? One approach planned by NASA is the StarLight mission, scheduled for launch in 2006. The Starlight mission is a "technology pathfinder mission" in NASA's search for new worlds. The Starlight mission will combine the light of two small formation flying space telescopes to create a 125-meter "virtual" space telescope or "sparse array" telescope – the largest ever flown in space at optical wavelengths. This is still a long way from 1000 m, but this is a start. Since the aperture of the telescope is the limiting factor in resolving power, the Starlight mission will use two telescopes that will appear as to sections of a larger optic. The two beams will then be combined into one image. The distance between the separate telescopes making up the interferometer must be very precisely known and controlled. StarLight's autonomous formation flying technologies will control the distance between the two spacecraft to within 10 centimeters (4 inches). This is a daunting guidance and controls task. Not only must the separation distance between the two telescopes be controlled, but the "flatness" of the simulated larger optic must also be controlled to within less than microns. At this time, nobody has a good handle on flatness control over such large distances.

MASINT

Measurement and Signature Intelligence (MASINT) is technically derived intelligence that detects, locates, tracks, identifies, and describes the unique characteristics of fixed and dynamic target sources. It is sort of an all encompassing field for intelligence gathering of signals that are emitted at the edge of the envelope of detectability and understanding. SETI actually should fall under MASINT collection.

MASINT is considered highly reliable since it collects performance data and characteristics on actual targets that do not intend to create an indication of presence or activity. In the event of an alien invasion, we would be forced to rely heavily on MASINT to uncover details of the alien invasion force technologies.

MASINT is predominately collected by dedicated technical sensors and processed/exploited by the MASINT production infrastructure. However, "MASINT is also derived from additional processing and exploitation of data sometimes collected by sensors of the other collection disciplines". Once again, MASINT is an all encompassing area of measurement and scientific study.

As part of the *Intelligence Community of the 21st Century* study (IC21), the House of Representatives of the One Hundred Fourth Congress Permanent Select Committee on Intelligence reviewed the MASINT discipline for its relevance in the Intelligence Community's (IC) future. The study suggested that the future in terms of specific technologies the Community will have to face is extremely unpredictable. And that decision was reached about Earth technologies. Imagine how much more difficult alien technologies would make MASINT collection.

The study's major findings include:

- MASINT can provide specific weapon system identifications, chemical compositions and material content and a potential adversary's ability to employ these weapons.
- MASINT, as a specific and unique discipline, is not well understood by both the IC and user communities. Therefore, the potential of its future contributions may be limited.
- MASINT is both a true, unique collection/analysis discipline and a highly refined analytical technique of the traditional disciplines.
- MASINT straddles strict disciplinary definitions.It may use collection techniques of, but does not fit neatly into any one or all of the more recognized "traditional" disciplines of IMINT, SIGINT, HUMINT, etc.

- MASINT is the least understood of the disciplines and is perceived as a "strategic" capability with limited "tactical" support capabilities. However, MASINT has a potential ability to provide real-time situation awareness and targeting not necessarily available from the more classic disciplines.
- MASINT is a science-intensive discipline that needs people/scientists well versed in the broad range of physical and electrical sciences. Such scientists cannot typically be professionally developed with the IC. They must come from academia fresh with scientific knowledge from experimentation and research. Nor can they continue to be "proficient" in their areas of expertise if they remain in government employ for an entire career.

Some of the study's major recommendations that we believe parallel the Sixth Column include:

- The MASINT technical management function should be contained within the construct of a multi-intelligence disciplined technical collection agency which oversees the coordinated employment of all technical collection systems.
- The IC should create a "U.S. MASINT System" analogous to USSS and USIS.
- The IC must be able to tap into any/all U.S. resources, including those not specifically within the IC, that have the ability to input into intelligence data bases. This includes having better access to, and guidance of, national laboratories.
- The IC needs a budgeting mechanism that is equivalent of "ready cash." This would provide the ability to readily fund fleeting or promising technologies, R&D efforts (without penalty for those technologies/or scientific breakthroughs that do not bear fruit), or unplanned operational opportunities. This authority needs to be analogous to a venture capitalist.
- The IC needs to examine the feasibility of pursuing trial personnel management programs that provide incentives to recruit and maintain the necessary scientific experts.

Another significant review done in preparation for future intelligence needs is the The National Intelligence Council project, NIC 2020 PROJECT. The review was done to help uncover the most important influences that will shape our world to the year 2020. This ambitious, yearlong study engaged a broad range of foreign and domestic experts who were challenged to think in new and provocative ways about the forces that will drive global developments.

The study was not meant to be an exercise in prediction or crystal ball gazing; our project may be called "2020" but neither our hindsight nor our foresight is 20/20. Rather, after considering the forces that will shape our world, a variety of regional and global scenarios were developed that represent alternative futures — and these futures will be discussed and debated through a series of conferences, in papers, and through dynamic Web-based interactions.

We believe that the types of consideration and preparation discussed in IC 21 and NIC 2020 demonstrate the level of preparation the Sixth Column desires. But we need to perform similar studies taking the possibility of alien invasion into account.

Special Tactics and Strategies

In the event of an alien invasion we will most definitely have to employ "out of the box" tactics and strategies in order to defeat them – perhaps simply to survive. Standard military doctrine should be considered as the first line of tactics and strategies, but new innovative concepts on how to interact, engage, and defeat the ET invader must be studied.

We suggest that a team within the *Sixth Column* would make a collection of all known literature, movies, television shows, cartoons, comic books, and websites about or involving alien interaction. Studying these storylines would accomplish several things:

1. An example basis would be created giving the team reference concepts. An example would be to say that a particular alien is like the Borg from *Star Trek:TNG* or like the "bugs" in *Alien* or *Starship Troopers*. Having this reference knowledge would make the study move faster and create a standard nomenclature.
2. A catalog of scenarios and solutions would be created. The science fiction database would allow us to track ideas that have been considered and then to analyze those ideas from a real life standpoint. For example: could we really fly a commandeered alien fighter into the mothership, hack into the alien central computer network, and upload a virus that turns their shields off as was done in *Independence Day*.
3. Seeing story after the story of "out of the box" concepts would soon change the definition of "out of the box" to the team. Studying the science fiction concepts in detail might enable the team to realize how to actually accomplish some of them sooner than the natural evolution of technology. This will raise the bar for what the *Sixth Column* truly believes is "out of the box".
4. The constant exposure to science fiction alien invasion scenarios would desensitize the *Sixth Column* from the initial shock if the invasion occurred and would enable faster response to the scenario.

Modeling and Simulation

The team that makes science fiction study and analysis part of their mission would also be the team that works the combat and warfare modeling and simulation. There are numerous combat and war gaming models and simulation tools used by our military strategist and military future technologist today. The simple coupled differential equations based on the Lanchester model shown in Chapter 2 is a basic example. There are much more complicated statistically based models available.

There are also a few tools used widely by the defense community that run "approved scenarios" for combat simulation. The two tools for force-on-force modeling that are accepted to date by DoD are JANUS, and CASTFOREM.

JANUS is a high-resolution, stochastic, two-sided, interactive, computerized ground combat simulation. It was developed in the 1970s by Lawrence Livermore National Laboratories for force-on-force simulation. It is used for combat developments, doctrine analysis, tactics investigation, scenario development, field test simulation, and training.

CASTFOREM is the Combined Arms and Support Task Force Evaluation Model. The tool is a highly robust simulation environment that can model individual entities (weapons, troops, vehicles) at resolutions required to address the study issues of interest for that particular simulation run. It is a brigade force-on-force, closed-loop stochastic combat model comprised of and captures output data for:

- Command and Control (C2)
- Communications
- Combat Service Support (CSS)
- Engineering
- Surveillance
- Engagements
- Maneuvers
- System/Environment

There are other models for individual aspects of warfighting simulation such as helicopter combat, fightercraft missions, bombers, UAVs, armored vehicles, and more. An effort is being made to day by the DoD to allow inputting these other models as modules for the JANUS and/or CASTFOREM environments.

The missile defense arena has also developed many models for simulating the National Missile Defense. Some of these simulations even have hardware-in-the-loop capabilities. In fact, there have even been missile launches combined with radar simulations running together nearly at real-time.

The *Sixth Column* would need to make a significant effort to create out of existing tools a means of simulating combat scenarios with theoretical technologies on both sides of the combat. The Earth defense systems could be simulated in battles against alien forces of various capabilities with such a tool. These simulations might indeed lead us to ideas, concepts, scenarios, tactics, and strategies that could give humanity a chance for survival. The simulations might also show us where we have technology gaps and would alert us to areas of development that need more funding.

It is also a great match to have the science fiction scenario study group combined with the simulation group. As the effort to study science fiction examples leads to possible alien invasion scenarios they can be implanted in the combat modeling and simulation tools. This effort would be critical in developing capabilities that would allow humanity to survive an alien invasion.

Wolves in Sheep's Clothing

One of the best tactical advantages in force-on-force scenarios is ambush. The element of surprise can be a major force multiplier. The *Sixth Column* should investigate concepts that on the surface may appear to be harmless to the alien forces but might pack an extreme knockout punch. We will discuss one possible example here.

Depending on where the alien forces locate themselves this tactic may or may not be useful. But, consider the *Independence Day* scenario where alien saucers hovered only a few kilometers above the cities they intended to destroy. We suggest that lighter-than-air craft might be of use. Helium filled latex weather balloons can lift a considerable payload mass and at the same time they have very little radar cross section (RCS).

It is possible that a WMD could be attached to such a vehicle and it be lifted into the region of the spacecraft. The vehicle might not even be detected by the aliens if they depend on radar. It might even be that their sensors are designed to detect fast moving vehicles, rocket plumes, heat signatures from engines, and high energy weapons powered up. All of the things that would make missiles and fighter planes ineffective might be what would make a low technology solution like an airship successful.

John Ringo's *Legend of the Aldenata* books (Posleen wars) involved alien invaders whose technology detected motion at high momentum and did not assume slower flying things were a threat. In that series blimps were used to transport logistics equipment and support troops. Ringo suggested a low technology approach to circumvent an alien technology.

If the alien invasion ships were to be at twenty to fifty kilometers high then high altitude balloons might still be useful. High altitude weather balloons can reach these altitudes but they have never been steered to date. New concepts for steering them at altitude should be investigated.

The U.S. Air Force Research Laboratory Space Vehicles Directorate's Space Integration and Demonstration Division has a Near Space Vehicle Access Program. Near space is the region of the atmosphere between twenty and fifty kilometers. In this region there are presently no flight vehicles with the capabilities fly around in this region. The U2 and SR71 spyplanes can reach the lowest end of this and so can the Global Hawk UAV. Rockets and missiles can fly through this region but cannot stop until they are nearly fifty kilometers above it or further. At this point rockets can reach a low Earth orbit but are way out of the near space region. If aliens discovered this, that region might prove to be a sweet spot to hide from Earth technologies in plane sight. That is assuming that they are invulnerable to missiles.

The Near Space Vehicle Access Program is an effort to test and study balloons and aerostats that potentially are ideal vehicles for space environment qualification; meteorological measurements; optical, infrared, ultraviolet, and radar surveillance; radio and laser communications; and target simulation. A shortfall as compared to an alien invasion is delivering WMD packages with them. Of course this is probably against some treaty, although the use of near space is still up to some debate where most current treaties are concerned because we have not made use of this region before. Figure 6.11 shows a high-altitude balloon test at the AFRL Near Space Access Program test sight at Kirtland Air Force Base, New Mexico.

Figure 6.11 High altitude balloon test at AFRL Near Space Access Program flight test. Credit: AFRL Space Vehicles Directorate

Another effort to develop near space is the U.S. Army's Space and Missile Defense Command and the Missile Defense Agnecy's High Altitude Airship (HAA) program. This program is designed to develop technologies that will enable a high altitude airship or near space vehicle that has propulsion capabilities and will allow it to fly around in near space as well as maintain station over a location. The missile defense purpose for the system is to enable platforms that can hover over the United States and act as early sentinels for an incoming missile threat. Figure 6.12 shows the HAA concept term demonstrator design architecture.

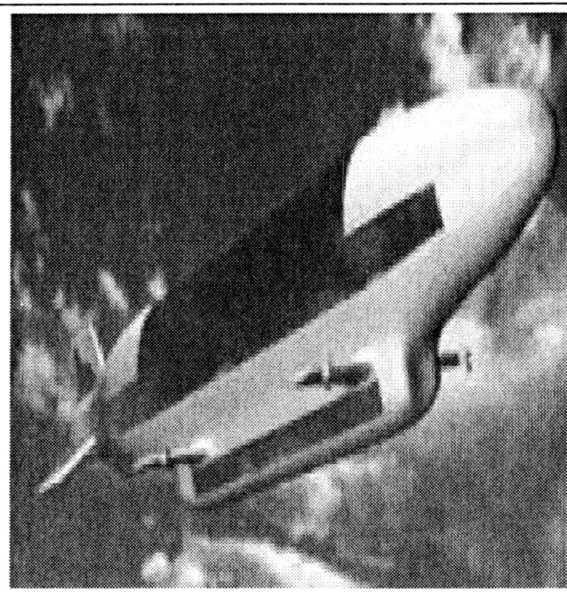

- Station-keeping Endurance—1 month
- Station-keeping Altitude—65,000 ft mean-sea-level (MSL)
- Payload Weight—500 lb
- Payload Power—3 kW
- Cruise Speed—25 kts
- Station-keeping Accuracy— < 2 km 50 percent of time, <150 km 95 percent of time
- Command and Control—Remotely Piloted

Figure 6.12 High Altitude Airship program's near term demonstrator architecture. Credit: U.S. Army SMDC HAA Program

Neither of these programs are considering carrying an offensive payload. However, the demonstrator for the HAA program has the goal of carrying a 500 lb payload. This would be a large enough capability to carry several small WMDs and potentially larger ones. Could this type of vehicle be a wolf in sheep's clothing that the ET invaders might overlook as not a threat? Other similar concepts should be investigated.

Fighter Planes Revisited

The section above describes the region of our atmosphere called "near space". We discussed there how we have no vehicles to fly in this region of space. We also have very limited capabilities above and below that region. We have no spacefighters and our air breathing fighters are limited to altitudes on the order of 15-20 kilometers. If the aliens parked their ships at geosynchronous orbits (35,000 kilometers) or higher we would have no means of attacking them.

We need a development effort for a spacefighter plane that has an envelope reaching from the ground to space. These fighters would enable us to attack the aliens away from Earth rather than us being forced to wait until they come down to destroy our cities. If they do not have to drop to low altitudes to destroy our cities then we might not even be able to engage them.

Of course, the technology is not available for a spacefighter presently. However, we need to be initiating research and development efforts that will enable them in the nearest possible future.

6.7 Chapter Summary

In this chapter we discuss the need for a program now that is developing defense strategies, tactics, and technologies that might enable us to survive (and maybe even win) in the event of an alien invasion. We discuss the need for being prepared in Section 6.1 and in Section 6.2 we explain why we should keep that preparedness a secret.

In Sections 6.3 and 6.4 we give a notional organization structure for this preparedness program as well as a notional budget for starting and operating it. For less than the minimal NASA budget an organization could be put in place and operated that would better prepare humanity for the possibilities of threats from outer space.

Section 6.5 discusses the responsibilities of the individuals that might work at this preparation organization. We suggest implementing the existing rules given in national security documents and protocols to the letter and spirit of them and no further rules need be created. The rules currently in place for highly sensitive and compartmented programs are such that would protect the public from "conspiracies" within the deeply classified alien invasion preparation program – or Planetary Defense Program.

Section 6.6 gives some discussion about various technologies that might be useful, if developed further, against ET invaders. Several technologies are discussed but are by no means exhaustive.

7

Conclusions and Discussion

What We Hoped to Accomplish With This Book

The idea for this book came about while the authors were working in a "brainstorming group" to develop solutions to the hard targets of today's military and intelligence gathering communities. During a break at the brainstorming session Dr. Taylor, Dr. Boan, Mr. Anding, and Dr. Powell brought up the idea that a truly hard problem would be to defeat an alien invasion. The analogy to asymmetric warfare came up and the four of us started seriously considering the possibilities.

Dr. Taylor and Dr. Boan spent the next few months outlining the book and through several discussions and investigations into the Drake equation and Fermi's Paradox began to realize that there was a significant possibility that such an invasion could occur.

Dr. Taylor was a panel member on several science fiction panels at science fiction conventions over then next couple of years where the topic of ET life and visitation were discussed. In almost all of the conferences the "mainstream" science guests all used the Drake equation analyses that existed to date and Fermi's Paradox as though they were proven to be fact they way they stood. Dr. Taylor took it on as a personal study effort to really understand these two analyses.

Dr. Boan and Mr. Anding began conducting technology assessments and trying to find where gaps might be that we should be looking into in the future. At the same time, Dr. Powell was looking into how we had handled the concept of alien invasion in the past by looking into how we have handled the UFO sighting phenomenon.

We brought these pieces together and realized that we needed a common language to speak about alien invasion scenarios. It was at this point that we decided using the science fiction stories would be a useful tool. The scenarios existed and we would not have to reinvent or name them. Also, the general public is more likely to be aware of Klingons and Romulans than a generic simulation name. We suggest this is a good approach for any future efforts.

So, we spent the next four years integrating this book into a cohesive story. There might still be some gaps here and there but our plan was to get this available to the world as soon as we could tell a decent story. The defense community needs to seriously be considering the points brought up in this book. And the civilian community needs to consider that alien invasion is possible and there is little or nothing humanity could do about it.

So, we originally set forth a number of goals for this text in the Introduction. We have made serious attempts to address each of these goals separately throughout the book. We have addressed each of these goals throughout this book to the degree that new ideas and open discussion is the next step. Open discourse and debate within the various scientific, military, political, civil, and academic communities will offer new insight and direction for achieving the goals for Planetary Defense.

As far as why there is a need for a Planetary Defense action plan there are over four hundred billion star systems in our galaxy alone. There are countless galaxies in the universe each with similar and even more star systems than our galaxy. There is a finite probability of intelligent life outside of our own star system. If we are alone there are vast quantities of resources available to us in those other systems. Man's tenure on Earth is resource limited. Sooner or later, we will need to venture out into the stars to find those other resources. It is likely that any other civilization would need to do the same. What if another civilization showed up at Earth to collect our resources? Some might argue that the probability of this happening is not worth considering. This is how we decided to start the book – explaining how likely an invasion of Earth is.

In Chapter 1 we brought up here an alternate (more realistic) analysis of the Drake Equation. We suggest that we have to consider ranges since the equation is a probability equation, probabilistic error must be considered. We hardly ever see that in the literature about the topic.

We also showed that Fermi's Paradox is really no paradox at all. The idea of it is based on a mathematical model that does not fit the real world. The fact that Fermi's Paradox has survived as long as it has is quite alarming to us. The community has blindly accepted the explanation for humanity being alone in the universe without giving the theory a truly rigorous analysis. Fermi's Paradox is simply put, wrong. It could possibly turn out to be the answer (that humanity is alone in the universe) but it would be merely a lucky outcome, because the math behind Fermi's analysis has no fit to the way things really evolve in the universe.

We also discussed possible scenarios for an invasion and we tried to give some examples that might lead to eventually categorizing alien types. We introduced how these types of aliens should follow statistical distributions and the Central Limit Theorem.

It should be obvious that we would like any ET that came to Earth to be benevolent. It would be great if ET brought cures for all diseases known to man, a cure for old age, and an *Encyclopedia Galactica*. It would be wonderful if they brought to us new technologies that would enable us to increase the efficiency of the internal combustion engine. This alone would postpone the impending fossil fuel crisis by many years. The ETs might even bring to us technologies that use other forms of energy that occur naturally in the universe such as tornadoes, hurricanes, geothermal, earthquakes, gravitational tides, vacuum energy fluctuations, or whatever. All of these benevolent instances would be welcomed. However, the Central Limit Theorem analysis suggests that there should be just as many bad aliens as there are good ones. So, instead of waiting for the good ones to come save us from ourselves and our unknown future, perhaps, we should prepare for the bad ETs visit.

In Chapter 2 we talked about warfare, the components of warfare, and how to go about understanding force-on-force interactions. We gave some simple simulations that suggested humanity's best hope is to hold out long enough to create an asymmetric situation with the alien force much like in the Soviet-Afghanistan war in the 1980s. Humanity would need to survive using tactics like the Mujahideen did against the Soviets. Eventually the Soviets decided that Afghanistan was no longer worth the cost and effort to maintain a campaign there. Perhaps we could do the same against the aliens.

We also discussed the potential energy capabilities of the aliens and what types of defensive/offensive capabilities we would have to implement in order to combat them. High energy weapons of WMD scale seem to be the best hope. New smaller systems need to be developed and unique delivery capabilities need to be considered.

We spent some time in Chapter 3 describing in detail possible alien motivations based on science fiction examples. This section would be a good starting point for future classification possibilities. We identified some possible personality types.

Chapter 4 talks about the "need to know" and why or why not some things should be classified. Even among the four authors there was some disagreement as to when or how the public would need to know about alien invaders. This chapter should offer plenty of "food for thought" and will perhaps lead to some interesting future debate.

Chapters 5 and 6 are more along the lines of what we should do in the event of an invasion and what we should be doing before an invasion occurs. Our best hope for survival is to begin preparing now. If the invasion never occurs then we will have sparked a major technological development period in our history, which would most likely also spark an economic boom. Being prepared would also give us better chances of survival against things like meteor impacts, mass epidemics, major natural disasters, and many other potentially large-scale hazardous scenarios. As it stands now, humanity is ill prepared for major natural hazards. In fact, we typically react or respond and our means of civil preparation is simply considering means of reacting. We are ill prepared and should be considering our options more seriously.

So in the end, what did we hope to accomplish with this book? We at least hoped to spark some interesting debate about the topics discussed within it. As of now, the topic is typically left for the science fiction enthusiast community. We hope that this book might shift discussion of the topic at least a few millimeters closer to the mainstream arena.

We hope that the civil, defense, and intelligence communities will take the book seriously and consider the ideas discussed within. If it turned out that a *Sixth Column* were created and humanity seriously began preparing for major events such as NEO impact and alien invasion, then we would say we were truly successful.

But what is most likely is that at least some people will be reached with this book. We hope that upon reading it that some points of agreement and disagreement will spark debate and maybe even future action. We hope that future efforts will allow us to expand on the book with future editions but that will be years down the road as it took us more than five years to get this one together.

We have tried to relay a message with this book in that we hope to express the philosophy of preparedness. Preparedness is the only chance that humankind has for survival in the event of an invasion from some Extra-Terrestrial source. Whether the source of the invasion is merely a Near Earth Object hurtling toward an impact with us or if it is Klingons off the port bow, if we wait until the threat shows up it will be far too late to do anything about.

Development efforts take years and even decades. If we plan to have a launch system that will enable us to intercept a NEO before it reaches near Earth space, then we had better start designing and building and testing the thing now.

If we suspect a greater threat from an alien civilization, we must have a war machine that can defend our civilization and enable our survival. We are presently not developing such capabilities at any appreciable rate. The reason is that most members of the general public do not perceive there to be any threat. The perception was discussed in detail in the first chapter of this book and we hope that we have relayed the fact that the probability of invasion is finite. There is no way of knowing if the finite probability is very small or very large but it is definitely finite. Any scientist who tries to convince the public that the probability of such an invasion is infinitesimally small or zero is wrong and most importantly irresponsible.

Even if the probability of an ET invasion is very small but finite, then we should be prepared. The probability of people on Earth being killed by tornados is fairly small and most people do not prepare for them. It is guaranteed, however, that if you are in a tornado shelter that is properly designed and prepared during a tornado incident that you will survive.

The cost of invasion preparation is not unbelievable; in fact it is on the order of the NASA budget, which is tiny, compared to other parts of the national budget. Most people do not realize that the NASA budget is diminishingly small. The price of not preparing for such an invasion is extinction!

We as a species should desire to survive and thrive in the universe. It is to our benefit to survive and thrive. Therefore it is to our benefit to prepare to survive rather than hope for survival through luck or divine intervention. We must be prepared.

We hope that this book leaves you with some thought provoking questions. We hope that it will stimulate more discussion as to the possibilities that might face us in the future for more reasons than good (or bad) science fiction scenarios. The possibilities are real, however small or large they may be, so we cannot simply ignore them.

There is a saying that Heinlein used a few times that goes something like, *"I'd rather have a gun and not need it than need a gun and not have it."*

The ideology behind this statement *is* the reason for this book. So to paraphrase or add corollary to Heinlein, "*We had rather be prepared for an alien invasion and it never happen than not be prepared for an invasion and be faced with one!"*

Bibliography

"Brazilian Official Report on the Trindade UFO." Fate 18,3 (March 1965): 38-48.
"New Evidence on IGY Photos." The A.P.R..0. Bulletin (January 1965): 1,3-8.
"UFO Photo Certified by Brazilian Navy Labeled a Hoax by USAF." The UF0 Investigator 1,10 (July/August 1960): 3.
"UFOs in Latin America." In Hilary Evans with John Spencer, eds. UFOs 1947-1987:
Adams, Douglas. *Hitchhikers Guide to the Galaxy*
Alcubierre, M. Class. Quantum Grav. **11**, L73 (1994), *gr-qc/0009013.*
an exchange with Moore, Charles B., Todd, Robert, Rodeghier, Mark and Randle, Kevin. *Project Mogul and the Roswell Crash, International UFO Reporter.* (March/April 1995)
Asimov, Isaac. *Foundation Trilogy*
Assistant Director of Central Intelligence for Analysis and Production at the IC Colloquium. New Mexico State University. 15 September 2003
Barter, Neville, ed. *TRW Space Data*
Battle of the Planets (cartoon)
Beyond the Thunderdome (movie)
Brown, Charles. *Spacecraft Mission Design*
Cameron, James. *Strange Days* (movie)
Campbell, Bruce A. and McCandless, Samuel W. *Introduction to Space Sciences and Spacecraft Applications*
Carey, Thomas. "The Continuing Search for the Roswell Archaeologists: Closing the Circle", *International UFO Reporter.* (January/February 1994)
Clarke, Arthur C. *2001: A Space Odyssey*
COMETA Report
Craft, Michael. *Alien Impact*
D.C. Comics. *The Green Lantern*
DARPA Budget 2001 - 2003
Darwin, Charles. *Origin of Species.*
Director of Central Intelligence Directives (DCID's)
Dunnigan, James F. *How to Make War.*
Earth vs. The Flying Saucers (movie)
Evolution (movie)
Federal Acquisition Regulations

Fontes, Olavo T. *"The UFO Sightings at the Island of Trindade." The A.P.R.O. Bulletin Pt. I* (January 1960): 5-9; Pt. II (March 1960): 5-8; Pt. III (May 1960): 4-8.
Galganski, Robert. *The Roswell Debris: A Quantitative Evaluation of the Project Mogul Hypothesis, International UFO Reporter,* (March/April 1995)
Godwin, Parke. *Waiting for the Galactic Bus*
Goodman, Joseph. *Introduction to Fourier Optics.* 1998
Hall, Peter J. CSIRO Australia National Telescope Facility, Epping, NSW 1710
Hall, Richard H. Letter to Donald H. Menzel (November 2,1959). Hall, Richard H., ed. *The UFO Evidence.* Washington, DC: National Investigations Committee on Aerial Phenomena, 1964.
Heinlein, Robert A. *Orphans of the Sky*
Heinlein, Robert A. *Sixth Column*
Heinlein, Robert A. *Starship Troopers*
Heinlein, Robert A. *The Puppet Masters*
Heinlein, Robert A. *The Puppet Masters*
Hewes, Hayden. *"The Mystery Disk over Trindade Island."* UFO Report 7,1 (February 1979): 18-19,58.
Hopf, John T. *"Exclusive IGY Photo Analysis."* The A.P.*R.0.* Bulletin (May 1960): 1,4.
Hubbard, L. Ron. *Battlefield Earth*
Huyghe, Patrick. *Fièld Guide to Extraterrestrials*
I Come in Peace (movie)
IC21: The Intelligence Community in the 21st Century. Staff Study, Permanent Select Committee on Intelligence, House of Representatives, One Hundred Fourth Congress
Independence Day (movie)
infocenter@cufos.org.
Invasion of the Body Snatchers (movie)
Kuhn, Thomas. *The Structure of Scientific Revolution*
Lee, Stan. *The Fantastic Four*
Lodders, Katharina and Fegley, Bruce, Jr. *The Planetary Scientist's Companion*
Lorenzen, Coral E. *The Great Flying Saucer Hoax: The UFO Facts and Their Interpretation.* New York: William-Frederick Press, 1962. Revised edition as Flying Saucers: The Startling Evidence of the Invasion from Outer Space. New York: New American Library, 1966.
Lovelace, Lt. Gen. James J. Jr. & Votel, Brig. Gen. Joseph L. The Asymmetric Warfare Group: Closing the Capability Gaps. www.army.mil
Lucas, George. *Star Wars* (movies)
Mad Max (movie)
Mallove, Eugene and Matloff, Gregory. *The Starflight Handbook: a Pioneer's Guide to Interstellar Travel*
Mars Attacks (movie)
Matloff, Gregory. *Deep Space Probes: to the Outer Solar System and Beyond*
McKinney, Jack. Robotech (series of books)
McKinney, Jack. The Blackhole Travel Agency

Men in Black (movie)
Menzel, Donald H. Letter to Richard H. Hall (November 27, 1959). *Menzel, Donald H., and Lyle G. Boyd. The World of Flying Saucers:A Scientific Examination of a Major Myth of the Space Age.*Garden City, NY: Doubleday and Company, 1963.
Menzel, Donald H., and Ernest H. Taves. *The UFO Enigma: The Definitive Explanation of the UFO Phenomenon.* Garden City, NY: Doubleday and Company, 1977.
Millis, Marc. www.grc.nasa.gov/WWW/bpp/
NASA Budget
National Industry Security Program Operating Manual (NISPOM)
NIC 2020 PROJECT (National Intelligence Council)
Pflock, Karl T. *Roswell Inconvenient Facts and the Will to Believe*
Predator (movie)
Project Greenglow BAE SYSTEMS
Project Grudge FOIA documents
Project Saucer FOIA documents
Project Sign FOIA documents
Randle, Kevin and Schmitt, Donald. *When and Where did the Roswell Object Crash?, International UFO Reporter.* (January/February 1994)
Randle, Kevin. *The Project Mogul Flights and Roswell, International UFO Reporter.* (November/December 1994)
Ringo, John and Taylor, Travis S. *Von Neumann's War*
Ringo, John. *A Hymn Before Battle*
Ringo, John. *Gust Front*
Ringo, John. *Hell's Fare*
Ringo, John. *Into the Looking Glass*
Ringo, John. *When the Devil Dances*
Robotech (cartoon)
Rodeghier, Mark and Chesney, Mark. The Air Force Report on Roswell: An Absence of Evidence, International UFO Reporter. (September/October 1994)
Rodeghier, Mark and Chesney, Mark. *The Final(?) Air Force Report on Roswell, International UFO Reporter,* (Winter 1995)
Rodeghier, Mark and Chesney, Mark. *What the GAO Found: Nothing About Much Ado, International UFO Reporter,* (July/August 1995)
Sagan, Carl. *Contact*
Sagan, Carl. *Cosmos*
Schutz, B.F. *A First Course in General Relativity*
Science Daily
Serber, Robert. *The Los Alamos Primer: The First Lectures on How to Build an Atomic Bomb*
Shklovskii, I.S. and Sagan, Carl. *Intelligent Life in the Universe*

Signs (movie)

Smith, Willy. *"Trindade Revisited." International UFO Reporter 8,4* (July/August 1983): 3-5,14.
Species (movie)
Spielberg, Steven. *Taken* (television miniseries)
Star Trek (books, movies, and television series)
Star Trek: The Next Generation™ (movies and television series)
Stargate (movie)
Stargate SG-1 (television series)
Stern & Mitton 1999
Stern, Alan and Mitton, Jacqueline. *Pluto and Charon*
Stern, Roger. *The Death and Life of Superman*
Taylor, James G. Naval Post Graduate School, "Low Resolution Combat Modeling and Simulation."
Taylor, Travis S. *Advanced Solar Sail and Laser Sail Propulsion Concepts for Interstellar Space Travel*
Taylor, Travis S. and Powell, T. C. *Current Status of Metric Engineering with Implications for the Warp Drive*, Joint Propulsion Conference Proceedings 2003
Taylor, Travis S. *The Quantum Connection*
Taylor, Travis S. *Warp Speed*
The 9/11 Commission Report (Authorized Edition)
The 40-Year Search for an Explanation, 97-113. London: Fortean Tomes, 1987.
The Day the Earth Stood Still (movie)
The Hidden (movie)
The King James Version of The Holy Bible
The Roswell Report: Case Closed, James McAndrew, Headquarters United States Air Force, Washington, DC, 1997.
The Thing (movie)
The X-Files (movie)
The X-Files (television series)
They Live (movie)
Titan A.E. (movie)
Transformers ™ (cartoon)
V (television series)
Vallee, Jacques, *UFO's in Space: Anatomy of a Phenomenon*
Van Den Broeck, C. (2000) *gr-qc/9906050*
Virtual Prototyping of RF Weapons. Office of the National Security Space Architect , 71-432-1300.
Ward, Peter D. and Brownlee, Donald. *Rare Earth: Why Complex Life is Uncommon in the Universe*
Webb, Stephen. *Where is Everybody?*
Webster's Dictionary
Weekly World News
Wells, H.G. *War of the Worlds*

Wertz, James R. and Larson, Wiley J. eds. *Space Mission Analysis and Design*
Winchester, Simon; *Krakatoa*, Harper Collins Publishers Inc. NY, NY 2003

World Wide Web Addresses:

http.//archieve.ncsa.uiuc.edu
http.//csep10.phys.utk.edu
http://asc-india.org/seismic/pakistan.htm
http://en.wikipedia.org
http://news.bbc.co.uk
http://nuketesting.enviroweb.org
http://oak.cats.ohiou.edu
http://snebulos.mit.edu
http://tigger.cc.uic.edu
http://www.esd.ornl.gov/programs/SERDP/EcoModels/popmodel.html
http://www.fas.org
http://www.gnu.org/software/gnuradio/index.html
http://www.greys.mcmail.com/ufo.htm
http://www.zgram.net/masint/
www.cdi.com
www.emergency.com
www.geo.mtu.edu
www.globalsecurity.com
www.imagers.gsfc.nasa.gov
WWW.NAIC.COM
www.niac.usra.edu
www.nro.gov
www.tapr.org
www.theestimate.com
Zubrin, Robert. *The Case for Mars*

Index

T

U

V

W

X

Y

Z

Notes

Notes

Notes

Notes

Printed in the United States
63478LVS00003B/101-106

9 781581 124477